The Use of Recycled Materials to Promote Pavement Sustainability Performance

The Use of Recycled Materials to Promote Pavement Sustainability Performance

Editors

José Neves
Ana Cristina Freire

MDPI • Basel • Beijing • Wuhan • Barcelona • Belgrade • Manchester • Tokyo • Cluj • Tianjin

Editors
José Neves
CERIS, Department of Civil
Engineering, Architecture,
and Georesources, Instituto
Superior Técnico
Universidade de Lisboa
Lisboa
Portugal

Ana Cristina Freire
Transportation Department
National Laboratory for Civil
Engineering
Lisboa
Portugal

Editorial Office
MDPI
St. Alban-Anlage 66
4052 Basel, Switzerland

This is a reprint of articles from the Special Issue published online in the open access journal *Recycling* (ISSN 2313-4321) (available at: www.mdpi.com/journal/recycling/special_issues/ Pavement_Sustainability).

For citation purposes, cite each article independently as indicated on the article page online and as indicated below:

LastName, A.A.; LastName, B.B.; LastName, C.C. Article Title. *Journal Name* **Year**, *Volume Number*, Page Range.

ISBN 978-3-0365-3844-0 (Hbk)
ISBN 978-3-0365-3843-3 (PDF)

Contents

About the Editors

José Neves

José Neves holds a Ph.D. in Civil Engineering and is a professor at the Department of Civil Engineering, Architecture, and Georesources, Instituto Superior Técnico, Universidade de Lisboa, in Portugal. He is a researcher at the Civil Engineering Research and Innovation for Sustainability (CERIS). Over his career, he has participated in R&D projects related to recycled and bio-based pavement materials. He is the author of scientific publications, including papers in peer-reviewed journals, chapters of books, book editions, and communications in international conferences. His main research interests are highway and airfield infrastructures, transportation geotechnics, pavement design and construction, pavement materials, recycling, and sustainability.

Ana Cristina Freire

Ana Cristina Freire, Ph.D. in Civil Engineering, is an Assistant Researcher at National Laboratory for Civil Engineering (LNEC), Head of the Transport Infrastructures Unit and Coordinator of the Pavement and Materials for Transport Infrastructures Laboratory, in Portugal. She is LNEC's Research Coordinator at FEHRL (Forum of European Highway Research Laboratories), chair of National Technical Committee for Standardization CT 154 –Aggregates and RILEM Technical Group member (WMR - Valorisation of Waste and Secondary Materials for Roads).

She is the author of technical and scientific publications, namely papers peer-reviewed, chapters of books and books editions, and communications in national and international conferences. She has taken part in national and European R&D&I projects concerning the use of secondary materials in road construction (alternative, recycled or bio-based materials). Her major research interests are transportation infrastructures; pavement maintenance and rehabilitation; behaviour assessment of road and airport pavements; innovative pavement materials and construction technologies; circular economy and sustainability.

 *recycling* 

Editorial

Special Issue "The Use of Recycled Materials to Promote Pavement Sustainability Performance"

José Neves [1],* and **Ana Cristina Freire** [2],*

1 CERIS, Department of Civil Engineering, Architecture and Georesources, Instituto Superior Técnico, Universidade de Lisboa, Av. Rovisco Pais, 1049-001 Lisbon, Portugal
2 LNEC, National Laboratory for Civil Engineering, Av. do Brasil 101, 1700-066 Lisbon, Portugal
* Correspondence: jose.manuel.neves@tecnico.ulisboa.pt (J.N.); acfreire@lnec.pt (A.C.F.)

Citation: Neves, J.; Freire, A.C. Special Issue "The Use of Recycled Materials to Promote Pavement Sustainability Performance". *Recycling* **2022**, 7, 12. https://doi.org/10.3390/recycling7020012

Received: 8 February 2022
Accepted: 22 February 2022
Published: 23 February 2022

Publisher's Note: MDPI stays neutral with regard to jurisdictional claims in published maps and institutional affiliations.

1. Introduction

Recycling road pavement materials allows for a more sustainable use of raw materials and contributes to creating a circular economy. The entire life cycle of the pavement products, focusing on their design, promoting circular economy processes, and fostering sustainable consumption, aims to ensure that the resources used are kept in the economy for as long as possible. Carrying out recycling policies may significantly impact civil engineering activities, including the construction and exploitation of transport infrastructure. Pavement engineering can also contribute to successfully achieve the sustainable development goals proposed by the United Nations [1], through a global framework supported by sustainable production employing green technologies.

Implementing consumption and production patterns based on recycling and adopting an industrial symbiosis approach can promote sustainable urban development under a carbon-neutral economy through green technologies. Due to intensive research and practice, recycling has been used in road construction, maintenance, and rehabilitation in the last few decades. Recycling pavement materials prevents the extraction of non-renewable resources and minimizes waste production and landfilling. It can save energy and decrease greenhouse gas emissions, thereby reducing pollution. Recycling effectively helps to reduce environmental impacts and combat global climate change.

The purpose of this Special Issue was to collect and publish technical and research papers, including review papers, focusing on the recycling of road pavement materials to promote pavement sustainability performance. Ten papers were published in total covering the use of construction and demolition waste (reclaimed asphalt pavement, recycled concrete aggregate and glass) and industrial waste (plastic and slag). The application of recycled materials concerns bituminous mixtures, concrete mixtures, and non-traditional interlocking blocks or cobbles. The most relevant contributions of each paper are briefly described in the following sections. The papers involved thirty-four authors from eleven countries of Europe (Belgium, Finland, Italy, the Netherlands, and Portugal), Africa (Nigeria), Asia (Malaysia and Saudi Arabia), Australia, and South America (Brazil and Colombia).

2. Use of Construction and Demolition Waste

2.1. Reclaimed Asphalt Pavement

Bituminous pavement courses are designed to present adequate characteristics, in terms of safety and comfort, during its period of life. After this period, construction, maintenance, and rehabilitation operations must be performed and, as a result, very high amounts of reclaimed asphalt pavement (RAP) are usually produced.

Reclaimed asphalt pavement (RAP) is a 100% recycled material obtained from road maintenance and rehabilitation operations. After adequate processing, such as crushing and screening, RAP can present high-quality and well-graded aggregates coated by bituminous mastic, thus becoming a secondary raw material proper to replace bitumen and virgin aggregates [2,3]. However, RAP applications, without downgrading, with incorporation in similar applications, still face several barriers, due to some lack of confidence in RAP recycling in new bituminous mixtures. The common maximum RAP incorporation rates vary between 10% and 50%, and between 0% and 20% for wearing courses [3]. The incorporation of high rates of RAP in new bituminous mixtures is still a challenge to be overcome to minimize life cycle costs and environmental impacts.

Vandewalle at al. [2] developed a comparative analysis on a real road pavement section, in which the real applied solutions were compared to alternative ones and were combined with the incorporation of five RAP rates into new bituminous mixtures (0%, 25%, 50%, 75%, 100%), in production, construction, and rehabilitation activities. The life cycle assessment (LCA) methodology was applied, and the results have been expressed in four damage categories: human health, ecosystem quality, climate change, and resources, together with 15 impact factors. The results demonstrated that both recycled and multi-recycled bituminous mixtures led to a decrease in the environmental impact when RAP was reused once or multiple-times. The benefits are greater for higher RAP rates presenting an average decrease of 19%, 23%, 31%, and 33% in all the four impact categories, for a 25%, 50%, 75%, and 100% RAP rate incorporation [2].

Antunes et al. [3] studied high RAP incorporation rates into new bituminous mixtures for wearing courses based on their long-term mechanical behaviour, taking into consideration the RAP bitumen mobilization degree, the evaluation of the RAP fractioning and mixing conditions, and both the mechanical and long-term behaviour of RAP mixtures. The behaviour of high RAP mixtures (75%) and virgin bituminous mixtures was compared. A crude tall oil rejuvenator was used to promote bitumen mobilization. The simulation of the ageing that occurs during mixture production and in-service life was accomplished by short- and long-term oven ageing procedures. The laboratory tests for the mechanical assessment were performed. The RAP bitumen mobilisation degree was evaluated, and a mixing protocol was developed and validated.

As a major conclusion, it was referred that, in general, the high RAP mixtures presented equivalent or even improved behaviours when compared with virgin bituminous mixtures. The performance of the high RAP mixtures presented good results even after ageing, allowing to conclude that these can present good long-term performances.

2.2. Recycled Concrete Aggregate

Since the cement industry is a major contributor to greenhouse emissions on a world-wide level, alternative materials are studied for the partial or complete substitution of cement in concrete. Recycled concrete aggregate (RCA) obtained from the demolition of old reinforced concrete structures is one of the recycled materials that can be reused to produce concrete and thus reduce the negative environmental impact of cement production [4,5]. However, some barriers need to be overcome regarding the use of RCA, namely the low demand for these materials and the costumers' unwillingness to pay more for them [4]. Many studies have considered the partial or complete replacement of cement in concrete. The use of fly ash and other by-products from the energy and mineral industry as additional cementitious materials in cement has a significant potential for reducing the carbon footprint of concrete [5].

Katar et al. [4] evaluated the application of construction demolition waste produced in Riyadh to manufacture high-strength concrete. Self-compacting concrete with 100% natural aggregate and three replacement levels (25%, 50%, 75%) of RCA was produced. Fly ash and a superplasticizer were added to obtain the adequate properties of flowability and cohesion in fresh state mixtures. The authors evaluated both fresh and hardened properties of the mixes, and J-ring, v-funnel, and slump flow tests were performed. Compressive strength tests after seven, 14, and 28 days were performed. The results confirmed that RCA can produce concrete with a reasonable compressive strength, its use being acceptable for structural applications.

Rintala et al. [5] presented a case study as part of the EU-funded research project "Urban Infra Revolution" that estimated the cost prices of four different geopolymer concrete (cement-free binders) with different material compositions and carbon footprints, considering the raw material price fluctuation and the potential impact of carbon emission regulation through carbon price. Two major questions were presented: "What are the benefits of using the materials?" and "How much does it cost?". The authors concluded that the results seem to indicate that carbon pricing, at the actual rates, does not significantly change the cost-price difference between traditional and geopolymer concrete. This means that the cost-competitiveness of low-carbon concrete depends on the material mix type and the availability of critical side streams.

2.3. Glass

Glass waste is suitable for various applications, including in the cement and concrete industries, due to its pozzolanic properties being more intensive in fine-grained form. Megna et al. [6] described research on combining glass and marble wastes to produce a new sustainable mortar for non-structural pavement solutions. Based on the experimental characterization of different types of mortars, the authors have confirmed the pozzolanic properties of the glass waste that led to the production of a hydraulic binder suitable to replace the conventional cement in concrete production.

3. Use of Industrial Waste

3.1. Plastic

Plastic wastes are a major global environmental issue, and their recycling and reuse are becoming more and more investigated. The diversity of plastic properties is enormous, and different approaches can be adopted to incorporate plastic wastes into pavement materials. The types of plastic covered in the Special Issue are polyethylene terephthalate (PET) [7], high-density polyethylene (HDPE) [8], acrylonitrile butadiene styrene (ABS) [9], polystyrene polymers (PS) [9], and low-density recycled polyethylene (LDPE) [10]. The authors have studied plastic waste applications in bituminous mixtures [7], concrete mixtures [8], interlocking plastic blocks [9], and sand/recycled-plastic cobbles [10] for pavements of roads [7,8] and other trafficked areas (e.g., parking areas, sidewalks, bike paths) [9,10].

In general, bituminous mixtures are considered a promising application for plastic wastes to achieve more sustainable pavements. Plastic wastes are being addressed as modifier agents of bituminous binders or as substitutes of aggregates. Mashaan et al. [7] investigated the effect of PET from plastic bottles on modifying a bitumen binder to be used on a 14 mm dense-graded asphalt for wearing course, composed of granite aggregates and a 4.9% optimum binder content. The authors studied the rheological properties of the plastic modified bitumen and the mechanical properties of the plastic modified bituminous mixture. Improved stability and resistance to permanent deformation were observed, more significant for 8% PET by weight of the bituminous mixture.

Other pavement applications of plastic wastes were concrete mixtures [8] and non-conventional blocks or cobbles [9,10]. In these applications, plastic waste was used for partial or total replacement of natural aggregates. Tamrin and Nurdiana [8] studied the incorporation of HDPE lamellar particles from diverse origins in concrete mixes for non-structural pavement applications. The authors concluded that the concrete with a 10 MPa compressive strength had the best resistance to adding HDPE, and 5% and 5×20 mm were the optimal content and size, respectively. Gabriel et al. [9] investigated interlocking plastic pavers composed of 70% ABS and PS from electronic equipment (e.g., computers) and 30% of other polymers and residual materials (e.g., other plastic or metal wastes). The 100% recycled blocks resulted from shredding, agglutination, and pressing procedures performed at specific temperature conditions. The authors carried out laboratory tests that confirmed similar properties compared to traditional blocks made of concrete for light traffic conditions. Sanchez-Echeverri et al. [10] evaluated the use of LDPE from recycled plastic bags to manufacture cobbles with 10 cm × 20 cm × 4 cm dimensions. The cobbles were composed of 25% plastic and 75% sand. The experimental research performed in the laboratory demonstrated the adequacy of the cobbles for pedestrian and lightweight traffic pavements. In complement, those authors also presented a market study to implement a factory in Colombia to produce these recycled cobbles.

3.2. Slag

Slags, in particular steel slag, lead slag cooper slag, and tin slag, are some of the industrial waste materials that have been studied to validate their application as replacement of natural aggregates. Tin slag (TS) is an industrial waste that is accessible and still under-utilized. About 2 million tons of this waste is landfilled worldwide. Olukotun et al. [11] studied the use of TS as a substitute for fine aggregates in cement mortar, considering different percentages of incorporation (0%, 25%, 50%, 75%, 100%). Three water/cement ratios of 0.5, 0.55, and 0.6 were used to prepare the tested specimens. Laboratory evaluation was conducted at fresh and hardened states and after 3, 7, and 28 days of water-curing of the testing specimens. The workability and the mechanical properties of mortar specimens were evaluated. According to the results, the authors considered that TS could be applied as a substitute for natural sand to produce mortars, thus promoting a reduction in costs and in natural resource depletion and leading to the sustainability of natural fine aggregates.

4. Final Remarks and Future Trends

The guest editors believe that this group of ten papers, published in this Special Issue, gave a significant contribution to promote the circular economy through the pavement sustainability performance of recycled materials. Other studies, namely regarding the application and validation of different types of alternative raw materials into pavements, may also be published in the following Special Issue: "The Use of Recycled Materials to Promote Pavement Sustainability Performance II".

Funding: This research received no external funding.

Acknowledgments: The guest editors are grateful to the MDPI Editorial Office for the support during peer review and publication of the papers included in the Special Issue. We also thank the authors for deciding to disseminate their investigation in this Special Issue, and the reviewers for their collaboration in ensuring the desired quality of the papers.

Conflicts of Interest: The authors declare no conflict of interest.

References

1. United Nations. Transforming Our World: The 2030 Agenda for Sustainable Development. Available online: https://sdgs.un.org/2030agenda (accessed on 31 January 2022).
2. Vandewalle, D.; Antunes, V.; Neves, J.; Freire, A.C. Assessment of Eco-Friendly Pavement Construction and Maintenance Using Multi-Recycled RAP Mixtures. *Recycling* **2020**, *5*, 17. [CrossRef]
3. Antunes, V.; Neves, J.; Freire, A.C. Performance Assessment of Reclaimed Asphalt Pavement (RAP) in Road Surface Mixtures. *Recycling* **2021**, *6*, 32. [CrossRef]
4. Katar, I.; Ibrahim, Y.; Abdul Malik, M.; Khahro, S.H. Mechanical Properties of Concrete with Recycled Concrete Aggregate and Fly Ash. *Recycling* **2021**, *6*, 23. [CrossRef]
5. Rintala, A.; Havukainen, J.; Abdulkareem, M. Estimating the Cost-Competitiveness of Recycling-Based Geopolymer Concretes. *Recycling* **2021**, *6*, 46. [CrossRef]
6. Megna, B.; Badagliacco, D.; Sanfilippo, C.; Valenza, A. Physical and Mechanical Properties of Sustainable Hydraulic Mortar Based on Marble Slurry with Waste Glass. *Recycling* **2021**, *6*, 37. [CrossRef]
7. Mashaan, N.; Chegenizadeh, A.; Nikraz, H. Laboratory Properties of Waste PET Plastic-Modified Asphalt Mixes. *Recycling* **2021**, *6*, 49. [CrossRef]
8. Tamrin; Nurdiana, J. The Effect of Recycled HDPE Plastic Additions on Concrete Performance. *Recycling* **2021**, *6*, 18. [CrossRef]
9. Gabriel, L.T.; Bianchi, R.F.; Bernardes, A.T. Mechanical Property Assessment of Interlocking Plastic Pavers Manufactured from Electronic Industry Waste in Brazil. *Recycling* **2021**, *6*, 15. [CrossRef]

10.	Sanchez-Echeverri, L.A.; Tovar-Perilla, N.J.; Suarez-Puentes, J.G.; Bravo-Cervera, J.E.; Rojas-Parra, D.F. Mechanical and Market Study for Sand/Recycled-Plastic Cobbles in a Medium-Size Colombian City. *Recycling* **2021**, *6*, 17. [CrossRef]

11.	Olukotun, N.; Sam, A.R.M.; Lim, N.H.A.S.; Abdulkareem, M.; Mallum, I.; Adebisi, O. Mechanical Properties of Tin Slag Mortar. *Recycling* **2021**, *6*, 42. [CrossRef]

recycling

Article

Laboratory Properties of Waste PET Plastic-Modified Asphalt Mixes

Nuha Mashaan *, Amin Chegenizadeh * and Hamid Nikraz

Department of Civil Engineering, School of Civil and Mechanical Engineering, Curtin University, Perth 6102, Australia; h.nikraz@curtin.edu.au
* Correspondence: nuhasmashaan@postgrad.curtin.edu.au or nuhas.mashaan1@curtin.edu.au (N.M.); amin.chegenizadeh@curtin.edu.au (A.C.)

Abstract: Commercial polymers have been used in pavement modification for decades; however, a major drawback of these polymers is their high cost. Waste plastic polymers could be used as a sustainable and cost-effective additive for improving asphalt properties, attaining combined environmental–economic benefits. Since 2019, in Australia, trial segments of roads have been built using waste materials, including plastic, requiring that laboratory evaluations first be carried out. This study aims to examine and evaluate the effect of using a domestic waste plastic, polyethylene terephthalate (PET), in modifying C320 bitumen. The assessment of several contents of PET-modified bitumen is carried out in two phases: modified bitumen binders and modified asphalt mixtures. Dynamic shear rheometer (DSR) and rolling thin film oven tests (RTFOT) were utilised to investigate the engineering properties and visco-elastic behaviour of plastic-modified bitumen binders. For evaluating the engineering properties of the plastic-modified asphalt mixtures, the Marshall stability, Marshall flow, Marshall quotient and rutting tests were conducted. The results demonstrated that 6–8% is the ideal percentage of waste plastic proposed to amend and enhance the stiffness and elasticity behaviour of asphalt binders. Furthermore, the 8% waste PET-modified asphalt mixture showed the most improvement in stability and rutting resistance, as indicated by increased Marshal stability, increased Marshall quotient and decreased rut depth. Future fatigue and modulus stiffness tests on waste plastic-modified asphalt mixtures are suggested to further investigate the mechanical properties.

Keywords: asphalt; waste plastic; visco-elastic properties; Marshall stability; rutting resistance; environmental impact

Citation: Mashaan, N.; Chegenizadeh, A.; Nikraz, H. Laboratory Properties of Waste PET Plastic-Modified Asphalt Mixes. *Recycling* **2021**, *6*, 49. https:// doi.org/10.3390/recycling6030049

Academic Editors: José Neves, Ana Cristina Freire and Carlos Chastre

Received: 17 May 2021
Accepted: 12 July 2021
Published: 14 July 2021

Publisher's Note: MDPI stays neutral with regard to jurisdictional claims in published maps and institutional affiliations.

1. Introduction

Rapid economic and industrial growth has increased Australia's waste production. In an attempt to manage these wastes and reduce their impact on the environment, different research projects have been conducted at Curtin University, Western Australia. These projects were focused on using waste materials to improve the building materials industry [1–3]. Results from these projects have shown the significance of utilising waste materials in building work and, in particular, improving geotechnical solutions and asphalt pavement reinforcement.

On the 1 January 2018, China enforced a ban on importing plastic waste, followed by similar prohibitions in other countries, such as India and Malaysia. These bans have had a significant impact on Australia's waste recycling industry [4,5]. The report stated that during 2018, the annual consumption of plastic was over three million tonnes, with only 9% recycled. It can be seen that during the last 17 years, Australia faced an extraordinary increase in waste materials as a result of people's daily lifestyle demands [4]. The consequent plastic waste produced has increased hazards and pollution [4,5]. As a result, the recycling of plastic waste in an environmentally friendly method is of great interest.

Using waste materials instead of new materials in the construction of roads has two significant benefits: substantially reduced costs and reduced waste going to landfill. Along with the importance of these benefits, the future development of using waste plastic in bitumen modification must consider how it enhances the properties of the mixture [6–9]. The current paper and another published paper [9] both belong one big project of using domestic waste plastic in the improvement of roads in WA, Australia. The published paper is different from the current submitted paper, because it only focused on binder properties in short- and long-term aging using the pressure aging vessel (PAV) method. Plastic has long been used in the asphalt industry showing competitive properties with commercial elastomer polymers in terms of improving engineering properties, such as rutting resistance and stiffness properties [10–13].

However, the addition of waste inclusions like PET would possibly lead to heterogeneous binders with brittle characteristics and low resistance to thermal and fatigue cracking. Previous studies focus on virgin plastic polymers and pay less attention to recycled plastics. Other studies [14–16] have displayed positive results using plastic in asphalt modification; nevertheless, few argue the use of a significant amount of plastics, such as PET and high-density polyethylene (HDPE), for improving the mechanical properties and durability of modified asphalt. Durability is the ability to resist deformation in the long-term service of asphalt life. Of interest, these studies [11–16] reported that adding plastic polymers could notably improve rutting resistance [15,16]. Studies also confirmed that adding plastic enhanced the workability and stability of the mixture [13,14,16]. Further studies have indicated that using waste plastic also results in improved rutting resistance of asphalt mixtures [16–18].

According to previous studies [12,17–20], a waste plastic content of 4% is suggested as the ideal in asphalt to achieve good properties in term of strength, stability, stiffness, better durability and rutting resistance. On the other hand, studies by [21–24] suggested 6% waste plastic content is essential for an enhanced modifier that could increase the fatigue life and cracking resistance of asphalt. The application of recycled plastic to improve pavements' properties has been carried out and evaluated in several countries, such as in the United Kingdom, Canada, India, the Netherlands and New Zealand, over the last seven years. In 2012, the city of Vancouver, Canada, used plastic waste as an alternative additive for reinforcing warm-mix asphalt [25]. As reported, three trail sections in Vancouver used local waste plastic in a 19 mm Superpave surface course warm-mix asphalt, thus helping to reduce the impact of greenhouse gases and improve air quality with 20% savings in energy used during mixing. Another example of the significance of utilising waste plastic in the pavement industry comes from the Netherlands in 2015 [26]. According to the construction company, a road fashioned out of recycled plastic would be able to resist low temperatures of −40 °C and up to highs of 80 °C, and be anti-corrosion and long-lasting for up to 50 years after construction [26].

According to investigations' reports [26–33], one of the effective ways of modifying asphalt is waste plastic, which would also be a way of supporting the environment and ecosystems. In addition, utilizing plastic polymers in asphalt potentially enhances the bitumen's temperature susceptibility and stiffness; this enhancement of bitumen results in an enhancement in the rutting and fatigue cracking resistance of asphalt pavement.

Lately, a valuable application of plastic waste has been tested in New Zealand to modify asphalt mixes [34]. According to their report, a large-scale trial of asphalt made with recycled plastic was conducted using 250 tonnes of plastic containers that would otherwise have been sent to landfill. Despite a few field trials in Brisbane, Melbourne and Sydney, since 2019 [35] no documented investigation has been reported in Australia. An earlier study by [35] was conducted using UK commercial plastic waste products added to C320 bitumen. Although some performance indicators were tested, the study [35] did not investigate the effect of local waste PET plastic on the elasticity and engineering properties of C320 asphalt using the wet-mix method. Consequently, there is a vital need to examine the performance of PET plastic-modified asphalt. This study aims to investigate and evaluate the impact of waste plastic in improving the engineering properties of asphalt mixes. The study will examine and evaluate the impact of waste plastic-modified asphalt on the enhancement of ageing stiffness performance and rutting resistance of the mixtures.

2. Materials and Methods

2.1. Materials

C320 bitumen binder was used in this study, supplied by SAMI Bitumen Technologies, Perth, Western Australia. Table 1 illustrates the physical properties of C320 bitumen. Local waste plastic bottles (PET) were collected, washed and used as a bitumen modifier after grinding to a size of 0.45 mm. Figure 1 shows the materials used in this research study. Typical, 14 mm, dense-graded asphalt for course surfacing was used. Granite aggregate was used, which is the most common natural aggregate in Western Australia. Table 2 displays the physical properties of the aggregate. Table 3 illustrates the particle size distribution of the aggregate.

Table 1. Physical properties of C320 binder.

Property	Value	Units	Methods/Standards
Viscosity at 60 °C	320	Pa.s	AS2 341.2
Viscosity at 135 °C	0.5	Pa.s	AS 2341.2
Penetration at 25 °C	40	0.1 mm	AS 2341.12
Flashpoint	250	°C	AS 2341.14

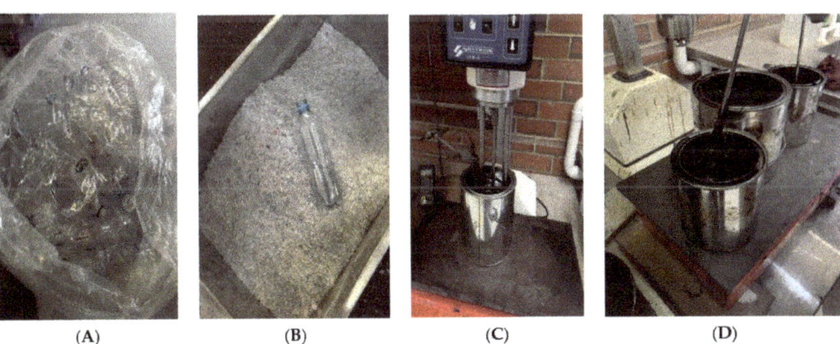

(A)   (B)   (C)   (D)

Figure 1. (**A**) Materials used in this research study of waste PET plastic; (**B**) plastic after grinding; (**C**) binder mixer using C320 bitumen; and (**D**) PET-modified C320 binders.

Table 2. Physical properties of the aggregate.

Property of Course Aggregate	Standard	Value/Limits
Water absorption (%)	AS 1141.6.1	0.4 < 2
LA value (%)	AS 1141.23	24.3
Aggregates crushed ACV (%)	AS 1141.21	23.8
Apparent particle density (g/cm^3)	AS 1141.6.1	2.692
Particle density: A dry basis (g/cm^3)	AS 1141.6.1	2.663
Particle density: A SSD basis (g/cm^3)	AS1141.6.1	2.674
Property of fine aggregate		
Water absorption (%)	AS 1141.5	0.6
Apparent particle density (g/cm^3)	AS 1141.5	2.697
Particle density: A dry basis (g/cm^3)	AS 1141.5	2.633
Particle density: A SSD basis (g/cm^3)	AS1141.5	2.657

Table 3. Particle size distribution of the aggregate.

Sieve Size (mm)	Lower Limit	Upper Limit	Selected Gradation
19	100	100	100
13.2	90	100	93
9.5	72	83	77
6.7	54	71	62.5
4.75	43	61	53.5
2.36	28	45	35.5
1.18	19	35	28.5
0.6	13	27	20.5
0.3	9	20	14
0.15	6	13	8.5
0.075	4	7	5

2.2. Samples Fabrication and Testing Methods

Sample preparation was conducted in two stages: plastic-modified bitumen samples and plastic-modified asphalt mixture samples. In the first stage, a high shear mixer was used to prepare the samples of waste PET-modified bitumen containing 0%, 4%, 6% and 8% by weight of bitumen. After several trial mixes in a laboratory, the authors selected the ideal mix conditions of 180 °C, 40 min and 2000 rpm for temperature, time and mixing velocity, respectively. The DSR tests of modified and un-modified binders were carried out to determine the visco-elastic behaviour, stiffness and elasticity and to investigate the rutting resistance before and after ageing. To assess the temperature susceptibility of the waste plastic-modified binder, a broad range of temperatures, of 50 °C, 58 °C, 60 °C, 64 °C and 70 °C, were assessed and recorded.

The second stage was to add the plastic-modified bitumen into the aggregate to prepare the plastic-asphalt mixtures by following the wet-mix method, and using the Marshall method. An optimum binder content of 4.9 was used in samples of plastic-modified asphalt with a plastic content of 4–8% by weight of mixes. In addition, 1.5% hydrated lime by weight of dry aggregate was used as the recommended filler by the Main Roads Western Australia (MRWA) standard. Marshall test (AS 2891.5-2004) stability, Marshall flow, Marshall quotient and wheel-tracking tests (AGP-T054-15) were conducted to better understand the impact of plastic on the performance properties of asphalt mixtures.

3. Results and Discussion

3.1. Results of Rheological Properties of Plastic-Modified Bitumen (before Ageing)

To assess the rutting resistance of the asphalt binders, the DSR test was conducted. According to the test results, all modified samples had acceptable and satisfactory values. As can be seen in Figures 2 and 3, waste PET-modified asphalt samples show large values of phase angle and complex shear modulus in comparison to the C320 binders (un-modified binders). In the specific high temperature range up to 70 °C, PET-modified asphalt samples showed an obvious improvement in terms of rutting resistance. The modified binder became less susceptible to deformation and, as such, improved the rutting resistance. The improvement in rutting resistance of the PET-modified samples through the different test temperatures showed better stiffness and elasticity performance in the modified samples. The results in Figures 2 and 3 display a better complex shear modulus and phase angle, which argues that PET particles integrate into asphaltenes, resulting in the swelling of the modified asphalt; this, in turn, leads to a significant improvement in binder viscosity and workability. This phenomenon indicates that the integration of the PET-asphalt blend leads to an improvement in elastic properties and is possibly associated with the characteristics and swelling of PET in the asphalt binder [6,8,36].

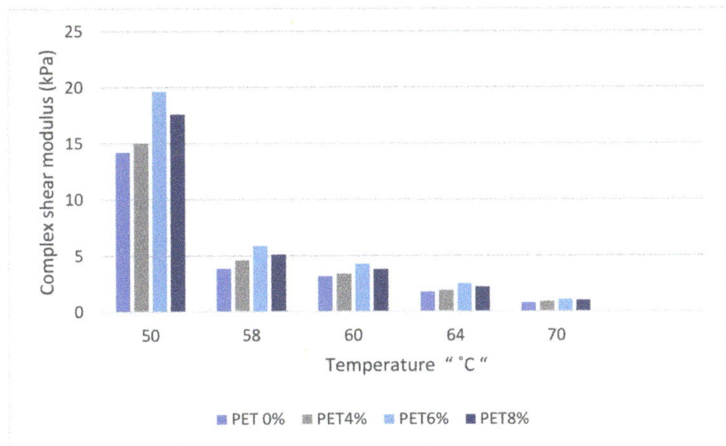

Figure 2. Complex shear modulus results of unaged PET-modified binders.

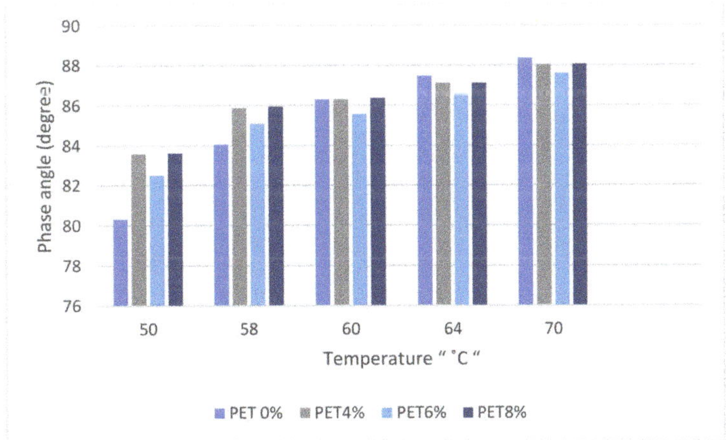

Figure 3. Phase angle results of unaged binders.

3.2. Results of Rheological Properties in Plastic-Modified Bitumen after RTFOT Ageing

Figures 4 and 5 show the impact of various PET contents on the visco-elastic behaviour of asphalt binders after RTFOT ageing. The DSR test assesses the visco-elastic properties, in terms of complex shear modulus and phase angle, through different testing temperatures from 50 °C to 70 °C. At all testing temperatures, the complex shear modulus and phase angle samples improved, as shown in Figures 4 and 5. The results demonstrate that the addition of PET into asphalt improves stiffness, giving better durability and ageing-resistance. This tendency was confirmed through the phase angle values, as shown in Figure 5. There was a noticeable decrease in the phase angle of the 6–8% PET-modified bitumen samples as compared to un-modified samples (C320 bitumen). This, in turn, indicates successful improvement of elasticity of the modified bitumen binder. As a consequence, the rutting resistance of PET-modified bitumen samples would be increased. As can be seen from Figures 4 and 5, the PET-modified binders would have less thermo-oxidative ageing compared to un-modified binders. In addition, modified binders show less hardening behaviour, better bitumen–aggregate bonding and thus, better resistance to deformation and cracking [8,35,36].

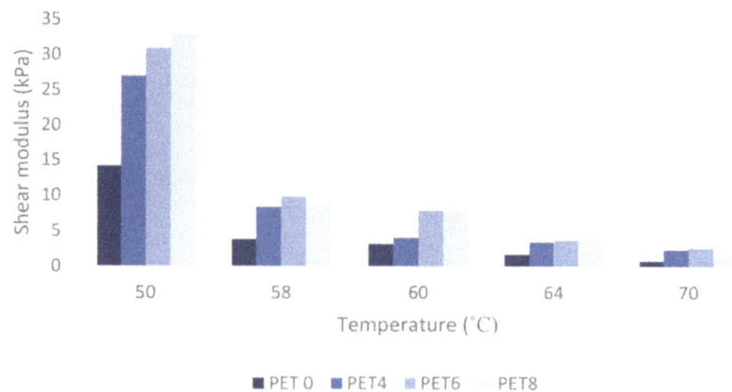

Figure 4. Complex shear modulus results of PET-modified binders after RTFOT.

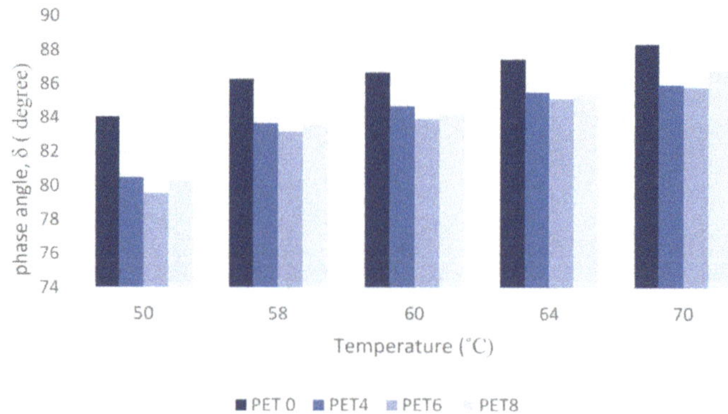

Figure 5. Phase angle results of PET-modified binders after RTFOT.

3.3. Marshall Test Results

The Marshall tests were simply conducted to investigate the ability of asphalt to withstand failure and rutting deformation under a substantial load of traffic. The values of Marshall stability, Marshal flow and Marshall quotient are displayed in Figures 6–8, respectively. As can be seen from Figure 6, the more waste plastic in the mixture, the higher the stability; therefore, the plastic-modified asphalt mixtures show better stability compared to the control mixture. The Marshall stability test is a well-known indicator of the resistance of bitumen materials to distortion, displacement, rutting and shear stress deformation [37]. The addition of plastic does result in a decrease in Marshall flow, as shown in Figure 7.

The Marshall quotient, referring to the ratio of Marshall stability to Marshall flow, can be used as a rutting resistance indicator; the results show that 6–8% PET plastic significantly increases the modified mixture's resistance to deformation. Consequently, the waste plastic polymer increases the asphalt's Marshall stability; however, it also shows a significant negative impact on Marshall flow, these results being similar to previous findings [23,35]. As can be seen in Figures 6 and 8, the increase in Marshall stability and Marshall quotient of the plastic-modified asphalt could be related to the greater dispersion of the PET in the bitumen during shear mixing, which could confer greater stiffness and result in higher stability. The higher Marshall quotient (MQ) values indicate the plastic-modified asphalt has become stiffer, and indicates that the modified asphalt is more resistant to permanent deformation [17]. The Marshall properties are considered direct indicators of pavement performance [35]; higher stability and lower flow are associated with better pavement performance. From the above results, we can expect that the asphalt mixtures modified with waste plastic would have improved performance properties. Figures 6–8 show higher correlation factors of $R^2 = 0.985$, $R^2 = 0.979$ and $R^2 = 0.988$ for stability, flow and MQ, respectively.

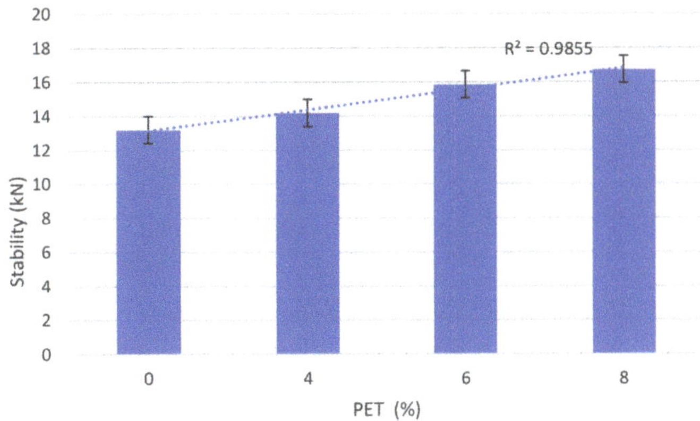

Figure 6. Marshall stability results.

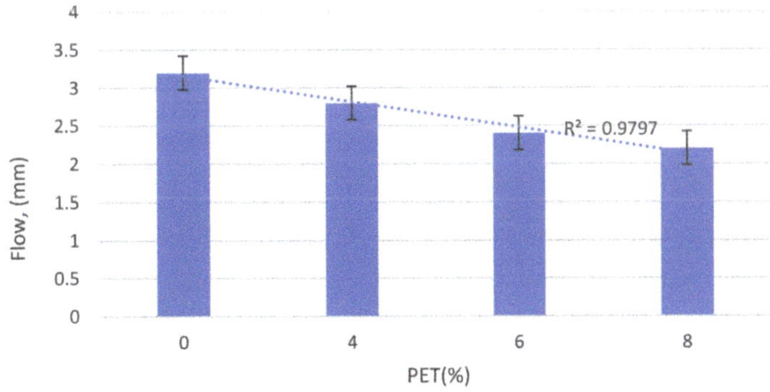

Figure 7. Marshall flow results.

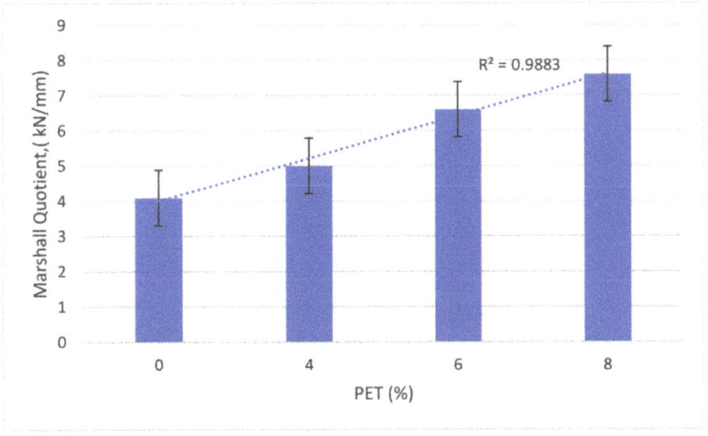

Figure 8. Marshall quotient results.

3.4. Wheel Tracking Tests Results

Rutting is one of the main distresses impacting road surfaces, especially in hot climates. Since rutting deformation is linked to the asphalt binder sensitivity to stresses and temperatures when using modified binders [9,21], it is vital to investigate the impact of the modified asphalt to clarify rutting resistance. The data of rutting tests used were the mean average of three readings for each PET content. As shown in Figure 9, the 4–8% waste-PET plastic modification reduced the rut depth of the modified mixtures compared to un-modified C320 (0% PET). The average rut depth of PET at 4, 6 and 8% was 8.82 mm, 5.59 mm and 3.25 mm, respectively. These results indicate that waste plastic would enhance the mixture's deformation resistance. The diversity of physical and chemical properties of PET plastic and C320 asphalt might justify the results of Figure 9. Specifically, the chemical properties are affected during the blending process of plastic-bitumen using the high shear mixer, which would lead to changes in the particle dimensions of PET plastic-bitumen-binders. Therefore, the engineering properties of PET-modified asphalt would be reinforced and improved in terms of elasticity increases. The increase in elasticity is due to the grouping of molecules and their bonds with each other [21,23]. Consequently, the waste PET-modified binders have the ability to improve the rutting resistance of pavements during their construction life and could be an effective additive in terms of reducing the costs of construction and maintenance.

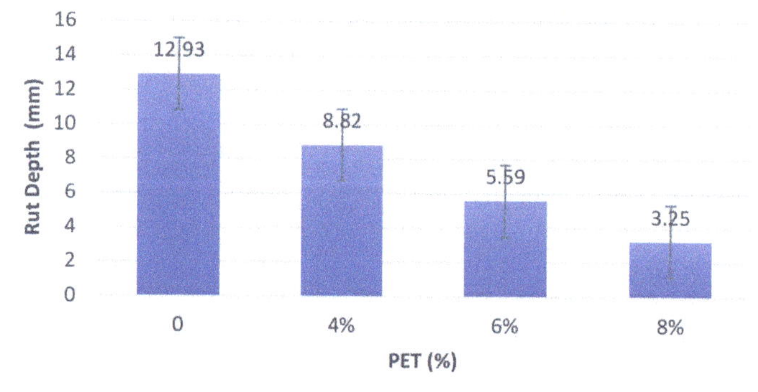

Figure 9. Rutting results vs. PET plastic content.

4. Conclusions

Bituminous pavements are vulnerable to cracking and rutting with certain temperature fluctuations; low temperatures can result in cracking; medium temperatures impose fatigue; and high temperatures result in a rutting problem. In line with such facts, changing and modifying the composition phase can enhance the engineering properties of the bituminous mixture through the injection of different additives. In the current study, waste plastic is used as a recycled additive to improve the engineering properties of bituminous mixtures through modification of the bitumen binder in the mix. Based on the current study, the notable findings are as below:

1. The study shows a promising and innovative approach of using local waste PET plastic as a modifier in C320 bitumen and asphalt mixtures.
2. According to the DSR results, samples of PET-modified asphalt indicate better performance in term of decreasing the asphalt's susceptibility to deformation at high temperatures, which would result in better rutting resistance. The complex shear modulus was increased and values of the phase angle reduced due to addition of PET to the binder. At the testing temperatures of 50–70 °C, a better rutting resistance was achieved which can be attributed to the elasticity of materials.

3. Based on short-term ageing (RTFOT) results, PET additives were less ageing, showed a higher complex shear modulus and had better elasticity with a better phase angle. These results indicate an ability to resist ageing during construction and offer better durability during long life service of asphalt pavement.
4. The waste PET-modified mixtures show good stiffness and improved stability, offering better resistance to shear stress under heavy loads. The values of the Marshall stability for all waste PET-modified asphalt mixtures were higher compared to the control mixture (C320); however, waste plastic had less impact on Marshall flow. The higher Marshall quotient (MQ) values indicate the plastic-modified asphalt became stiffer and indicates that the modified asphalt is more resistant to deformation.
5. The optimum content of waste plastic was 8%, resulting in smaller rut depth and improved resistance of modified mixtures to rutting deformation. Further stiffness and fatigue tests on asphalt mixtures samples are recommended to better understand the mechanical properties of waste PET plastic on modified C320 asphalt mixtures. Future tests should conduct various mixing conditions and quantify the reinforcement capability of PET-modified asphalt mixtures. In addition, possible future works could include the evaluation of PET contents variations over time, including aging simulation, alternated with UV radiation.

Author Contributions: Conceptualization, N.M. and A.C.; methodology, N.M. and A.C.; validation, N.M., A.C. and H.N.; formal analysis, N.M.; investigation, N.M.; resources, N.M. and A.C.; data curation, N.M.; writing—original draft preparation, N.M.; writing—review and editing, N.M.; visualization, N.M.; project leader, A.C.; project administration, N.M., A.C. and H.N. All authors have read and agreed to the published version of the manuscript.

Funding: This research was funded by the Australian Government Research Training Program RTP.

Institutional Review Board Statement: Not applicable.

Informed Consent Statement: Not applicable.

Data Availability Statement: The data presented in this study are available on request from the corresponding author.

Acknowledgments: Authors would like to acknowledge the contribution of the Australian Government Research Training Program RTP in supporting this research project.

Conflicts of Interest: The authors declare no conflict of interest.

Sample Availability: Not available.

References

1. Chegenizadeh, A.; Keramatikerman, M.; Panizza, S.; Nikraz, H. Effect of powder recycled tire on sulfate resistance of cemented clay. *J. Mater. Civ. Eng.* **2017**, *29*, 04017160. [CrossRef]
2. Chegenizadeh, A.; Keramatikerman, M.; Santa, G.D.; Nikraz, H. Influence of recycled tyre amendment on the mechanical behavior of soil-bentonite cut-off walls. *J. Clean. Prod.* **2018**, *177*, 507–515. [CrossRef]
3. Piromanski, B.; Chegenizadeh, A.; Mashaan, N.; Nikraz, H. Study on HDPE effect on rutting resistance of binder. *Buildings* **2020**, *10*, 156. [CrossRef]
4. O' Farrell, K. *2017–2018 Australian Plastics Recycling Survey, National Report*; Project Reference A21505; Australian Government: Canberra, Australia, 2019.
5. Chin, C.; Damen, P. *Viability of Using Recycled Plastics in Asphalt and Sprayed Sealing Applications*; Technical Report No. AP-T351-19; Austroad Ltd.: Sydney, Australia, 2019; ISBN 978-1-925854-51-0.
6. Dalhat, M.A.; Al-Abdul Wahhab, H.I. Performance of recycled plastic waste modified asphalt binder in Saudi Arabia. *Int. J. Pavement Eng.* **2017**, *18*, 349–357. [CrossRef]
7. Hamedi, G.H.; Hadizadeh Pirbasti, M.; Ranjbar Pirbasti, Z. Investigating the effect of using waste ultra-high-molecular-weight polyethylene on the fatigue life of asphalt mixture. *Period. Polytech. Civ. Eng.* **2020**, *64*, 1170–1180.
8. White, G. A Synthesis on the Effects of Two Commercial Recycled Plastics on the Properties of Bitumen and Asphalt. *Sustainability* **2020**, *12*, 8594. [CrossRef]
9. Mashaan, N.S.; Chegenizadeh, A.; Nikraz, H.; Rezagholilou, A. Investigating the engineering properties of asphalt binder modified with waste plastic polymer. *Ain Shams Eng. J.* **2021**, *12*, 1569–1574. [CrossRef]

10. Al-Hadidy, A.; Yi-Qui, T. Effect of polyethylene on life of flexible pavements. *Constr. Build. Mater.* **2009**, *23*, 1456–1464. [CrossRef]
11. Al-Hadidy, A.; Yi-Qiu, T. Mechanistic approach for polypropylene-modified flexible pavements. *Mater. Des.* **2009**, *30*, 1133–1140. [CrossRef]
12. Awwad, M.; Shbeeb, L. The use of polyethylene in hot asphalt mixtures. *Am. J. Appl. Sci.* **2007**, *4*, 390–396. [CrossRef]
13. Attaelmanan, M.; Feng, C.P.; Al-Hadidy, A. laboratory evaluation of HMA with high density polyethylene as a modifier. *Constr. Build. Mater.* **2011**, *25*, 2764–2770. [CrossRef]
14. Vansudevan, R.; Sekar, A.R.C.; Sundarakannan, B.; Velkennedy, R. A technique to dispose waste plastics in an ecofriendly way-Application in construction of flexible pavement. *Constr. Build. Mater.* **2012**, *28*, 311–320. [CrossRef]
15. Mahdi, F.; Khan, A.A.; Abbas, H. Physiochemical properties of polymer mortar composites using resins derived from post-consumer PET bottles. *Cem. Concr. Compos.* **2007**, *29*, 241–248. [CrossRef]
16. Costa, L.M.; Silva, H.M.; Peralta, J.; Oliveira, J.R. Using waste polymers as a reliable alternative for asphalt binder modification-Performance and morphological assessment. *Constr. Build. Mater.* **2019**, *198*, 237–244. [CrossRef]
17. Hınıslıoğlu, S.; Ağar, E. Use of waste high density polyethylene as bitumen modifier in asphalt concrete mix. *Mater. Lett.* **2004**, *58*, 267–271. [CrossRef]
18. Casey, D.; McNally, C.; Gibney, A.; Gilchrist, M.D. Development of a recycled polymer modified binder for use in stone mastic asphalt. *Resour. Conserv. Recycl.* **2008**, *52*, 1167–1174. [CrossRef]
19. Ahmad, L.A. Improvement of Marshall properties of the asphalt concrete mixtures using the polyethylene as additive. *Eng. Technol.* **2007**, *25*, 383–394.
20. Sara, F.; Silva, H.M.; Oliveira, J.R. Mechanical, surface and environmental evaluation of stone mastic asphalt mixtures with advanced asphalt binders using waste materials. *Road Mater. Pavement Des.* **2017**, *20*, 316–333.
21. Ameri, M.; Nasri, D. Perfromance properties of devulcanized waste PET modified asphalt mixtures. *Pet. Sci. Technol.* **2017**, *35*, 99–104. [CrossRef]
22. Yu, B.; Jiao, L.; Ni, F.; Yang, J. Evaluation of plastic-rubber asphalt: Engineering property and environmental concern. *Constr. Build. Mater.* **2014**, *71*, 416–424. [CrossRef]
23. Sojobi, A.O.; Nwobodo, S.E.; Aladegboye, O.J. Recycling of polyethylene terephthalate (PET) plastic bottle wastes in bituminous asphaltic concrete. *Cognet Eng.* **2016**, *3*, 1133480. [CrossRef]
24. Khan, I.M.; Kabir, S.; Alhussain, M.A.; Almansoor, F.F. Asphalt design using recycled plastic and crumb-rubber waste for sustainable pavement construction. *Procedia Eng.* **2016**, *145*, 1557–1564. [CrossRef]
25. Ridden, P. *The Streets of Vancouver Are Paved with Recycled Plastic*; New Atlas: Vancouver, Canada, 2012. Available online: http://newatlas.com/vancouver-recycled-plastic-warm-mix-asphalt/25254/ (accessed on 20 October 2020).
26. Saini, S. *Forget Asphalt: A European City Is Building a Road Made Entirely out of Recycled Plastic*; Business Insider: Amsterdam, The Netherlands, 2015. Available online: https://sg.finance.yahoo.com/news/forget-asphalt-european-city-building-163000280.html (accessed on 20 October 2020).
27. Li, R.; Leng, Z.; Yang, J.; Lu, G.; Huang, M.; Lan, J.; Zhang, H.; Bai, Y.; Dong, Z. Innovative application of waste polyethylene terephthalate (PET) derived additive as an antistripping agent for asphalt mixture: Experimental investigation and molecular dynamics simulation. *Fuel* **2021**, *300*, 121015. [CrossRef]
28. Esfandabad, A.S.; Motevalizadeh, S.M.; Sedghi, R.; Ayar, P.; Asgharzadeh, S.M. Fracture and mechanical properties of asphalt mixtures containing granular polyethylene terephthalate (PET). *Constr. Build. Mater.* **2020**, *259*, 120410. [CrossRef]
29. Ghabchi, R.; Dharmarathna, C.P.; Mihandoust, M. Feasibility of using micronized recycled Polyethylene Terephthalate (PET) as an asphalt binder additive: A laboratory study. *Constr. Build. Mater.* **2021**, *292*, 123377. [CrossRef]
30. Nizamuddin, S.; Jamal, M.; Santos, J.; Giustozzi, F. Recycling of low-value packaging films in bitumen blends: A grey-based multi criteria decision making approach considering a set of laboratory performance and environmental impact indicators. *Sci. Total. Environ.* **2021**, *778*, 146187. [CrossRef]
31. Nizamuddin, S.; Jamal, M.; Gravina, R.; Giustozzi, F. Recycled plastic as bitumen modifier: The role of recycled linear low-density polyethylene in the modification of physical, chemical and rheological properties of bitumen. *J. Clean. Prod.* **2020**, *266*, 121988. [CrossRef]
32. BintiJoohari, I.; Giustozzi, F. Chemical and high-temperature rheological properties of recycled plastics-polymer modified hybrid bitumen. *J. Clean. Prod.* **2020**, *276*, 123064.
33. Santos, J.; Pham, A.; Stasinopou, P.; Giustozzi, F. Recycling waste plastics in roads: A life-cycle assessment study using primary data. *Sci. Total. Environ.* **2021**, *751*, 141842. [CrossRef]
34. Parkes, R. Recycled Plastic Used in Airport Asphalt; Roads & Infrastructure Australia: Christchurch New Zealand 2018. Available online: http://roadsonline.com.au/recycled-plasticused-in-airport-asphalt/ (accessed on 20 October 2020).
35. White, G.; Magee, C. Laboratory Evaluation of Asphalt Containing Recycled Plastic as a Bitumen Extender and Modifier. *J. Traffic Transp. Eng.* **2019**, *7*, 218–235.
36. Leng, Z.; Sreeram, A.; Padhan, R.K.; Zhifei, T. Value-Added Application of Waste Pet Based Additives in Bituminous Mixtures Containing High Percentage of Reclaimed Asphalt Pavement (Rap). *J. Clean. Prod.* **2018**, *196*, 615–625. [CrossRef]
37. Mashaan, N.; Ali, A.H.; Koting, S.; Karim, M.R. Performance Evaluation of Crumb Rubber Modified Stone Mastic Asphalt Pavement in Malaysia. *Adv. Mater. Sci. Eng.* **2013**, *2013*, 304676. [CrossRef]

recycling

Article

Estimating the Cost-Competitiveness of Recycling-Based Geopolymer Concretes

Annastiina Rintala [1,*], Jouni Havukainen [2] and Mariam Abdulkareem [2]

1   Department of Industrial Management, LUT University, 53850 Lappeenranta, Finland
2   Department of Sustainability Science, LUT University, 53850 Lappeenranta, Finland;
     jouni.havukainen@lut.fi (J.H.); mariam.abdulkareem@lut.fi (M.A.)
*   Correspondence: annastiina.rintala@lut.fi

Abstract: The cement industry is a major contributor to greenhouse gas emissions on a global scale. Consequently, there has been an increasing interest, in both academia and business, in low-carbon concretes in which Ordinary Portland Cement (OPC) is partially or fully replaced with industrial side streams. However, the realization of the environmental benefits of such materials depends on how competitive they are in the construction market, where low costs are a major competitive factor. This is not straightforward, as many types of concretes exist. Raw material prices vary, and costs can be influenced by governmental regulations via carbon pricing. This study presents a case study estimating the cost prices of four different geopolymer concretes with different material compositions and carbon footprints, considering the raw material price variability and the potential impact of carbon emissions regulation (carbon price). The case study demonstrates how material mix cost comparisons can be made openly and systematically. The results imply that carbon pricing, at the rates currently applied, does not significantly change the cost price difference between traditional and geopolymer concretes. Instead, cost-competitiveness of low carbon concretes depends heavily on the material mix type and the availability of critical side streams.

Keywords: recycling; concrete; costs; carbon footprint; carbon price

Citation: Rintala, A.; Havukainen, J.; Abdulkareem, M. Estimating the Cost-Competitiveness of Recycling-Based Geopolymer Concretes. *Recycling* **2021**, *6*, 46. https://doi.org/10.3390/recycling6030046

Academic Editors: José Neves and Ana Cristina Freire

Received: 25 May 2021
Accepted: 2 July 2021
Published: 5 July 2021

Publisher's Note: MDPI stays neutral with regard to jurisdictional claims in published maps and institutional affiliations.

1. Introduction

Concrete production generates approximately 8.6% of all carbon dioxide emissions worldwide [1], and a considerable amount of this comprises cement production. Each tonne of Ordinary Portland Cement (OPC) produced requires 60–130 kg of fuel oil or its equivalent, depending on the cement variety and the process used, and about 110 kWh of electricity. This accounts for around 40% of the average 0.9 tonnes of CO_2 emissions per tonne of cement produced, while the rest are attributed to the calcination process, other manufacturing processes, and transportation [2].

Due to this environmental concern, many studies have been devoted to the partial or complete replacement of cement in concretes. Utilizing fly ash and other by-products of the energy and minerals industry as supplementary cementitious material in cement (or as a raw material in cement-free binders, referred to as geopolymers in the literature) holds a considerable potential in lowering the carbon footprint of concrete. It is often mentioned that geopolymer binders have been shown to potentially reduce carbon dioxide emissions associated with the manufacturing of cement by up to 80% [3].

However, the actual sustainability benefit of low-carbon materials depends on the scale to which their manufacture and use can be put into practice. Economic feasibility has been among the major factors hindering the scale-up of geopolymer concrete in the markets. Carbon pricing, an increasingly used form of regulation, is a potential means for promoting low-carbon solutions, and in principle, it could level the cost price difference between traditional and low-carbon concretes. This induces the following research question: What is the cost-competitiveness of geopolymer concrete, and how can it be assessed?

This study approaches this question by presenting a case study estimating the economic feasibility of four new recycling-based geopolymer concrete mixes. The approach includes gathering raw material price information, calculating price ranges for material mixes, and adding the impact of potential carbon pricing scenarios to the model using life cycle assessment (LCA) data. The model provided by this study can be used to support decision-making alongside material development to assess the cost-competitiveness of low-carbon concrete mixes.

1.1. Recycling-Based Low-Carbon Concretes

The CO_2 emissions and their mitigation approaches in the cement industry have been a subject of numerous studies [4,5], and according to the estimates published by Cembureau [6], the amount of CO_2 emitted from cement production has dropped from 783 kg/t (1990) to 667 kg/t (2017) and could be reduced to 571 kg/t by 2030 by making use of various means, such as energy efficiency solutions, low-carbon fuels, clinker substitution, and carbon capture and storage. This study focuses on clinker or cement substitution with side streams.

Overall, the carbon footprint-reducing effect of side stream usage is based on the fact that in LCA studies, it is an extensive practice to assume that waste does not carry environmental impacts from the previous life cycle phases. This suggests that waste's life cycle begins from waste generation, for example, when it is disposed into a bin. This is defined as the zero-burden approach [7]. Thus, the environmental impacts of the secondary raw materials, such as industrial side streams or waste, are allocated to the primary production chains (i.e., to the main products and by-products, not the waste; [8]).

Side streams can be used as supplementary cementitious materials (SCMs) with OPC, or as precursors in other binder types that are used in alkali-activated concretes. Alkali-activated concretes (AACs) are synthesized by reacting an alkali silicate/alkali hydroxide solution with an aluminosilicate powder and water [9]. SCMs, such as fly ash or ground granulated blast furnace slag (GGBS) with a high Si/Al ratio, can be used as source materials (aluminosilicate). Allthough in many cases, "alkali-activated material" may be a more correct term, the term "geopolymer concrete" is used generally in academic literature and in business for concretes that are produced without grounded clinker (burned limestone). The same side streams can be utilized in many cement/binder types; therefore, the suitable side streams can be referred to generally as binder precursors.

Aluminosilicates in geopolymer mixes comprise mainly slags, ashes, and metakaolin. The majority (65%) of geopolymer mix designs in studies are based on fly ash, followed by fly ash/slag (12%), metakaolin (10%), and slag (6%) [10]. Alkali activators generally comprise alkali hydroxides (e.g., NaOH, KOH) and alkali silicates (e.g., Na_2SiO_3, K_2SiO_3), individually or in combination.

Aggregates, another basic component of concrete, are traditionally sand and gravel. Generally, aggregates are not chemically involved in the structure's formation, but their packaging affects the resulting structure's properties. Therefore, it is imperative that the grain size distribution is correct. In addition to sand and gravel, sand-like and gravel-like substances may be used, for example, tailings [11], recycled concrete [12], granulated fly ash [13], and municipal waste incineration bottom slag [14]. Very small grain aggregates are called fillers, and some have structural strengthening properties. Most frequently, granular fillers are used for this purpose, such as special mineral sands or fumes and ground ceramic products [15].

Geopolymer concretes are known to have a longer service life than traditional OPC concrete, because geopolymer-based materials possess good mechanical strengths [16], a high resistance against chemical attack [17], and a high temperature resistance [18]. Notably, in the literature, geopolymer concrete is often compared against CEM I-type OPC concrete [19–22]; however, there are different concrete types that can be considered traditional, because they follow the cement standard.

Different cement classes defined in the standard are presented in Table 1. If it is assumed that 15% of concrete is cement, 5% water, and 80% recycled aggregates (a typical mix for 20 MPa compressive strength), it is possible to achieve a 94% recycling rate within the current cement standard.

Table 1. Classification of cement types according to cement standard, adapted from [23].

Cement Type	Description
CEM I Portland cement	Portland cement and up to 5% of minor additional constituents (the original OPC)
CEM II Portland composite cement	Portland cement with up to 35% of other SCM, such as ground limestone, fly ash, or GGBS
CEM III blast furnace cement	Portland cement with a higher percentage of blast furnace slag, up to 95%
CEM IV pozzolanic cement	Portland cement with up to 55% of selected pozzolanic constituents
CEM V composite cement	Portland cement blended with GGBS or fly ash and pozzolanic material

Side stream-based binder precursors can be divided into already recognized materials in cement standard, such as silica fume, coal fly ash, and blast furnace slag. Additionally, there are other slags and ashes suitable for geopolymer purpose but are unrecognized in cement standard. Examples are bauxite residual from alumina production—also known as red mud [24]—and waste from mining and processing copper ore [25]. Furthermore, some consumer wastes and separate fractions of demolition waste, such as scrap glass [26], ceramic waste [15], and mineral wool [27], have shown potential for geopolymer purpose. There are studies that have replaced commercial alkali activators with recycling-based material [28,29]; however, this has not been implemented on an industrial scale.

1.2. The Markets of Low-Carbon Concretes

Currently, geopolymer concrete utilization is limited in construction, and a few reasons have been found for this: (1) current material standards hinder the acceptance of geopolymers (2) the availability of industrial side streams varies by location, as do the volumes and qualities of side streams; and (3) the costs of geopolymer production easily exceed the costs of OPC production [3]. Furthermore, several studies have discussed additional barriers, such as the development of suitable admixtures, required operator skill at batch plants, capital intensive setup of processing facilities, lack of long-term durability data, particularly field performance, and the development of appropriate test methods [30].

Despite these barriers, there are various reported examples of the manufacturing of geopolymer concrete products around the world. The case studies are often pilots, but there are also companies currently manufacturing geopolymer concrete materials or products (Renca, Wagners, Kiran Global Chem, Zeobond, Geopolymer Solutions, GeoMITS, Alchemy Geopolymer Solutions, and SQAPE Technology).

Geopolymer concrete has been used as a surface material in roads and fields. Examples can be found in pedestrian routes and cycling paths, e.g., in the Netherlands [31] and Australia, as well as heavy traffic roads in Russia [32]; fields, such as container fields in Johannesburg and wind power parks in Loeriesfontein, South Africa; airport runways, e.g., during the Persian Gulf War (hybrid OPC/geopolymer concrete) [33] or in Australia [34]. Geopolymer concrete has also been applied in repairing similar structures [35].

Another application area is housebuilding and industrial construction. Multi-story houses have been built in Russia, Poland, and China, with earliest known examples dating back to the 70s [32]. In agriculture, geopolymer concrete has been used in silo building; it is considered a hygienic material [32]. The material has also been applied to parts of buildings, such as floors in industrial buildings [31], masonry tiles and roof tiles [32], and composite panels [36]. It can also be applied in mortar or underlayment and in coating material, e.g., for corrosion protection and fire protection [37] or waterproof pool coating [38].

Due to its waterproof properties and durability, one area of application is structures that are closely connected with water, such as piers [39] and boat ramps [40], water construction in general [41], fountain structures, sewage pipes and other sewage structures [32], and the repair of the such facilities [42].

Furthermore, geopolymers are used in earth construction, where the strength requirements are considerably lower than in building construction [43]. Overall, the geopolymer concretes compete on the same markets as the traditional concretes. Although there is some attention to special applications, such as refractory concrete or acid-proof concrete, there are also similar products in the markets that are manufactured with current cement standards.

1.3. Price-Estimations Concerning Low-Carbon Concretes

When compared with OPC concrete, geopolymer concrete is usually more expensive, but slightly cheaper in some cases. Habert and Ouellet-Plamondon [44] compared the economic allocation of one-part geopolymers and discovered that they couldachieve an 80% reduction in costs compared with OPC. Yand et al. [19] concluded that the cost of their one-part alkali-activated BFS foamed concrete was slightly higher, by 1.0–1.11 times, than that of OPC concrete. Chan, Thorpe, and Islam [20] estimated the material and manufacturing costs for over 20 years of fly ash-based geopolymer cement and OPC and concluded that geopolymers would be 18% more expensive. McLellan et al. [45] estimated the production costs of four typical Australian geopolymer pastes and concluded that in one case the manufacturing costs were 7% lower, but in other cases they were up to 39% higher compared with OPC. Vilamová and Piecha [21] estimated that the material cost of fly ash-based concrete is twice the cost of OPC concrete. Ostwal and Chitawadagi [22] estimated the production costs of GGBS and fly ash-based geopolymer concrete blocks to be higher than those of traditional cement concrete blocks.

Notably, the cost comparisons presented above are not generally comparable because they have not been standardly performed. There is variation in which costs are considered in the cost estimations, in the used mix designs, and in the prices of used ingredients. However, studies have provided many distinct geopolymer concrete mixes, among which 125 formulations of geopolymer-based materials have been reported in recent reviews [46–49]. The reported mix designs offer a basis for estimating cost price ranges for geopolymer concretes.

1.4. Carbon Pricing

To reduce and mitigate carbon emissions, some national and international authorities have adopted carbon emissions regulation practices to encourage polluters to reduce the volume of greenhouse gases they emit into the atmosphere. The method usually takes the form of either a carbon tax or a requirement to purchase permits to emit, generally known as carbon emissions trading or "cap and trade."

A carbon tax is a tax levied on the carbon content of fuels. The term carbon tax is also used to refer to a carbon dioxide equivalent tax. The principle of cap and trade is as follows: in the European Union Emissions Trading Scheme (EU ETS), the European Commission determines the maximum amount of emissions allowed for each trading period corresponding to the total pot of allowances. This total amount is distributed to the various member states, whose national emissions trading authorities allocate a country-specific quota to industrial and energy production plants. The total amount is smaller than the current European Union (EU) emissions. Thus, not all companies involved in emissions trading can continue producing emissions as before. It is therefore profitable for some companies to reduce their emissions, and thus sell the released allowances to others in need.

The EU ETS is the largest multi-national greenhouse gas emissions trading scheme in the world. Phase I began operation in January 2005, with all 15 EU member states participating. The program caps the amount of carbon dioxide that can be emitted from large installations with a net heat supply in excess of 20 MW, such as power plants and carbon intensive factories. Currently, there are around 600 plants in Finland covered by carbon emissions trading. By 2019, over 40 countries, including Finland, had implemented carbon taxes. In Finland, CO_2 tax is included in fuel tax.

1.5. Prices of Carbon Allowances and Carbon Taxes

Carbon's prices have varied over time and place. Figure 1 presents some examples of carbon's prices and provides a preliminary perspective of the potential price range for the carbon emissions. Currently (28 September 2020), the Swedish carbon tax represents the highest carbon price worldwide. Notably, prices are not necessarily comparable between carbon pricing initiatives due to differences in the number of sectors covered and allocation methods applied, specific exemptions, and different compensation methods.

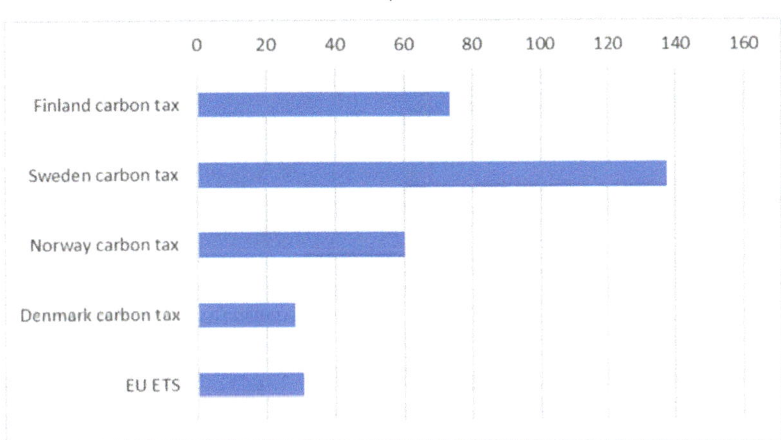

Figure 1. Examples of carbon prices [US$/tCO$_2$e], data from [50] (ETS: Emissions Trading System).

Despite the current policies and prices, it is also common practice—in American corporations—to calculate an "internal price on carbon." This internal price is calculated to assess the risk value of future projects when making economic investment decisions if the assets have a long lifespan and can be affected by energy policies in the future. The prices range from USD 6–7 per tonne of carbon dioxide at Microsoft to USD 60 per tonne at Exxon Mobil [51].

In the literature, carbon emission regulation has inspired many modeling studies that concern, for example, supply chain design and management [52–58]. The potential impact of carbon emissions regulation should also be noted in different contexts, including the construction context, but this has not yet received similar attention in the literature.

2. Results

This section presents the results of the material price search, cost-price estimates for concrete mixes, and analysis of the impact of carbon pricing on the cost-price differences between different material mixes. In addition, the generalizability of the cost estimation approach is discussed.

2.1. Estimates of Material Purchase Prices

On the Alibaba.com website, different suppliers report the price range of their products. The suppliers came from 36 countries, the majority from China (Figure 2). The price range may depend on both the purchase volume and the quality of the product. Figure 3 presents a summary of the price search. The number of options, N, mainly indicates the number of suppliers, but in some cases, the supplier sells the same material in different qualities, so it increases the number of options. The lower boundary of the price range was calculated as the average of the lower boundaries informed by the suppliers, and the upper boundary was calculated in an analogous manner.

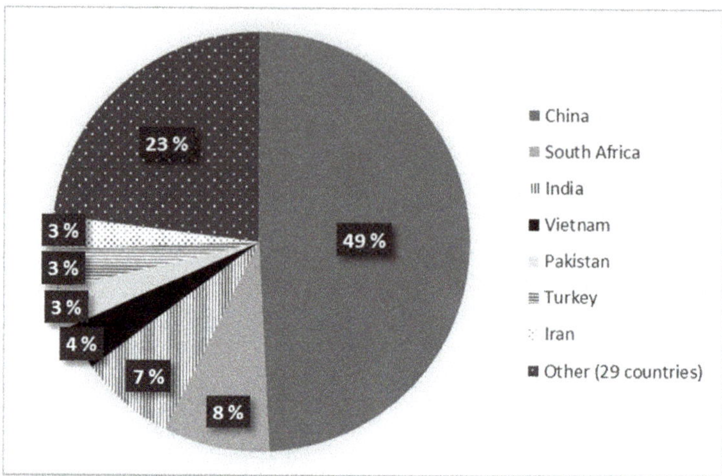

Figure 2. Concrete material supplier distribution by country in Alibaba.com, April 2019.

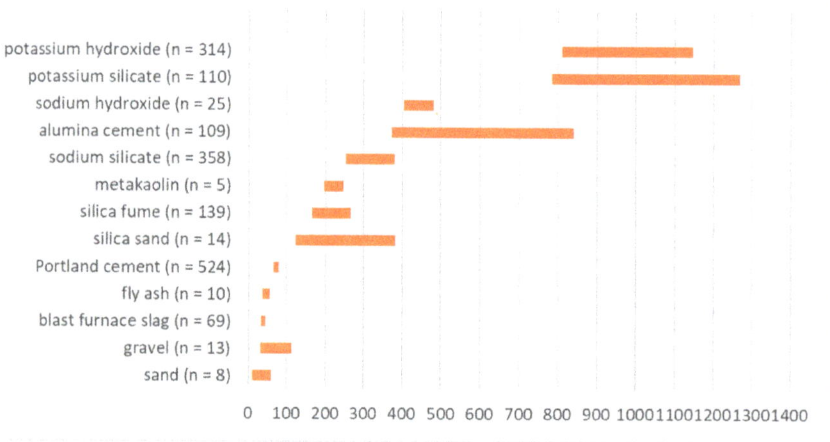

Figure 3. Average price ranges for the prices of different concrete raw materials in Alibaba.com in April 2019 [USD/t] n: number of options.

The price search focused on materials that were used in geopolymer concrete mixes in the case project, or in typical geopolymer concrete mixes presented in the literature. It should to be noted that all possible geopolymer concrete raw materials are not sold on open markets; therefore, the price information is not available.

For example, in the case of this project, the utilization of various local ashes and slags as aggregate was considered. Due to disposal fees, these side streams have a negative price in principle, but the required processing (e.g., drying and sieving) significantly increases their cost price. A large part of the processing costs consists of fixed costs; therefore, it is difficult to estimate the potential unit price for the processed material. Instead, a reference price was used, which was the price of the nearest commercial equivalent. This must be understood as a lowest possible price (range) for the material considered.

Notably, for Portland cement, the price range is relatively narrow despite the high number of suppliers; however, in the case of some ingredients of geopolymer concrete, especially for alkali activators (potassium/sodium hydroxide/silicate), the price range is wide. This is partly explained because these are sold in different forms (solid/liquid), grades, and in different purities. This makes it more difficult to estimate the actual possible prices of geopolymer concretes, whereas for traditional concrete, it is easier to estimate a reference price.

For gravel and sand, Alibaba.com does not seem to offer prices that are competitive with those offered by local suppliers in Finland. During the case project, such reference prices such as 5–10 €/m^3 for sand and gravel were discussed. However, the search results implied that in some other geographical areas, high-quality sand and gravel may have a higher selling price.

In this price search, it seemed that typical aluminosilicates—coal fly ash and blast-furnace slag—showed lower selling prices than Portland cement. All the suppliers of coal fly ash were from India, except one from Pakistan, and almost all suppliers of blast furnace slag were from China. It is difficult to ascertain only based on these data whether the quality of the ash and slag sold is suitable for binder manufacturing, but if so, their use as SCMs can lower the cost price of concrete. However, the relationship between demand and supply affects the selling price of the material. For example, in Finland, the prices of these materials are considerably higher; 110 EUR/t (134 USD/t) for blast furnace slag and 70 EUR/t (85 USD/t) for coal fly ash. In Finland, there is only one blast furnace, and coal fly ash has recently vanished from open markets due to poor availability. The open availability of critical side streams by location can be considerably estimated in studies that compare production and utilization volumes. For example, a considerable volume of unutilized crude steel slag has been noticed in China [59], and the same applies to coal fly ash [60]. Naturally, the utilization rates, which imply the prices, are not stable. Silica fume, a by-product of producing silicon metal or ferrosilicon alloys, is an example of a material that has turned from waste to a high-value product. Although once cheap, it is now an expensive high-performance cement supplement primarily used in bridges and other exposed structures that require high weathering performance [2].

In most concrete applications, the concrete material needs to be reinforced. In the reinforced geopolymer mixes presented in the literature [61–69], the amount of fiber is 19–156 kg/m^3 for steel fiber and 0.3–33 kg/m^3 for other fibers (glass fiber, polypropylene fiber, and polyvinyl alcohol fiber). The amount of reinforcement needed is assumed to depend heavily on the application of concrete. In the case project, the amount of fiber mixed was 3–4 kg/m^3 when the concrete was used for noise-barrier elements. Example prices for reinforcement materials are presented in Table 2. Although the prices USD/t are high, they do not have a significant effect on the cost price of the material if the quantity is as small as in the case.

Table 2. Price ranges for concrete reinforcement materials from Alibaba.com (USD/t).

Concrete Reinforcement Material	Price
Steel Fiber	698–1300 USD/t
Polypropylene fiber	1699–2699 USD/t
Glass fiber	540–1500 USD/t
Polyvinyl alcohol fiber	1999–2999 USD/t
Basalt fiber	2200–2800 USD/t

2.2. Cost Price Estimates for Example Mixes

Table 3 shows five mix compositions, and Table 4 shows their compressive strengths. Mix 0 represents a typical OPC concrete mix that can be used in construction. Other mixes are geopolymer concrete mixes. Mix 1 is a castable mix, and 2, 3, and 4 are 3D-printable mixes, all developed in the case project. For confidentiality, the mix designs are not revealed in more detail, but similar type (3D printable) geopolymer concrete mixes have been presented in the literature [70–72].

Table 3. Proportions of the different material mixes.

Constituent	Mix 0	Mix 1	Mix 2	Mix 3	Mix 4
Cement	27%				
Calcium aluminate cement				4%	
Activator		10%	15%	19%	0.3%
Waste precursor (CFA and GBFS)		25%	4%		37%
Metakaolin			9%	13%	
Fine aggregates	9%	13%	19%	13%	17%
Coarse aggregates	52%	45%	48%	43%	30%
Water	12%	6%	4%	6%	16%
Polypropylene fiber	0.14%	0.14%	0.14%	0.14%	0.14%

Table 4. Properties of the different material mixes.

Material Property	Mix 0	Mix 1	Mix 2	Mix 3	Mix 4
Compressive strength (28 days) [MPa]	50	25	37	30	14

Based on mix designs and material prices, it is possible to present price estimates for the mixes. However, there are some notable uncertainties. For materials that are rarely used in concrete mixes, the price information is not generally available. For example, in this case, recycled aggregates were used in all the mixes, but those aggregates do not have a market price, because they were produced only on laboratory scale. For simplicity, prices of all aggregates were assumed to be similar to sand and gravel. Additionally, the prices in Alibaba.com do not represent material prices in all locations. Therefore, the basic calculations can be used only for preliminary analysis.

Figure 4 presents a visualization of the price ranges of different concrete mixes. The lower boundary for the price range of a mix is simply calculated as the weighted average of the lower boundaries of material prices presented in the previous section. The upper boundary was calculated in an analogous manner.

The cost price range of Mix 4 is analogous to the cost price range of Mix 0 (OPC concrete). For Mix 1 and Mix 0, the ranges partly overlap. This means that under certain conditions, it is possible that the cost price of Mix 1 is closely similar to that of OPC concrete. For Mixes 2 and 3, the price ranges were clearly different when compared with Mix 0. Therefore, to compete in the markets, these mixes should offer clearly more value to the customers than the Mix 0.

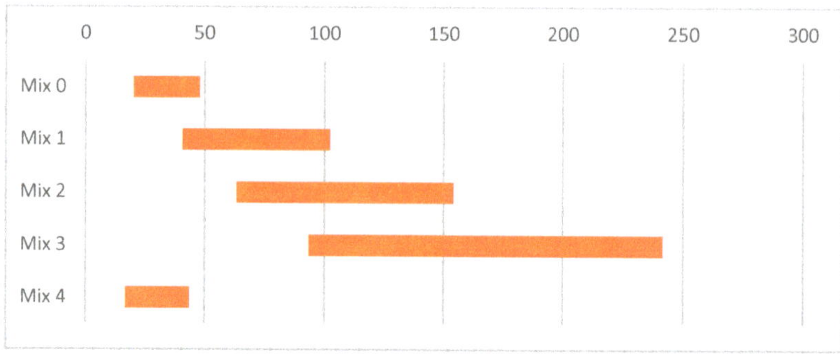

Figure 4. Cost price ranges for different material mixes [USD/t] (wet tonne).

This is a simplified preliminary cost price comparison, but it can be used in determining mixes that are by default clearly more expensive than the basic OPC mix. For example, the use of a potassium-based alkali activator, high activator proportion, and silica fume or silica sand anticipate a high-cost price for the mix. In the literature, numerous described mixes are designed for laboratory purposes and are not cost-optimized.

The comparison can be refined to account for the location where the mix is to be prepared. Potential freight can be added to the price estimates, and suppliers can be removed and added according to the possible delivery area. Activators' price ranges can also be narrowed by specifying the required activator characteristics in more detail. In the case project, for example, a higher price and/or freight would have to be applied to coal fly ash, which would clearly increase the price of Mix 4.

If no price information is available for the raw materials, the applicability of this procedure is limited. Notably, only material costs are considered here, while the actual cost price is also affected by manufacturing costs (mixing). The production of geopolymer concrete is somewhat more demanding than OPC concrete because dry materials (including aggregates) have to be dry-mixed, whereas in the production of OPC concrete, the moisture content of the aggregates can vary.

2.3. The Impact of Carbon Pricing on Cost Prices

Although it is often mentioned in the literature that geopolymers have the potential to offer 80% less carbon footprint than OPC concrete, the actual reduction depends on the mix and, in some cases, the carbon footprint of geopolymer concretes can even be higher than OPC concrete [73]. The carbon footprint reduction potentials of the case project mixes are presented in Table 5 based on the LCA performed in this project. The reduction potential is calculated for the mixes, both based on the LCA done in the project (Case 1) and expecting that CO_2 emissions of OPC concrete production will decrease by 25% (Case 2). Mix 4 had the highest reduction potential, and Mix 1 also held some potential even when the CO_2 emissions of OPC production decreased. The LCA was conducted assuming local material sourcing. If, for example, fly ash is transported between continents, emissions from transportation should also be taken into account in the calculation.

Table 5. Comparison of carbon footprints of different concrete mixes.

Measure	Mix 0	Mix 1	Mix 2	Mix 3	Mix 4
Carbon footprint [kg CO_2/m^3]	247	119	220	342	5
Carbon footprint [kg CO_2/t] *	107.20	51.61	95.64	148.79	2.08
reduction potential compared with OPC [1]		52%	11%	−39%	98%
reduction potential compared with OPC [2]		20%	−11%	−48%	55%

* Assuming 2,3 t/m^3 density; [1] current situation; [2] expecting 25% reduction in the CO_2 emissions of OPC concrete production.

Figure 5 shows how carbon pricing affects the material cost prices. This calculation assumes that carbon pricing is applied to the total CO_2 emissions of the material production. In this case, carbon pricing affects the cost differences of the mixes but does not change the order of the mixes. Mix 4 has a lower cost price than OPC concrete; (however, in the case of Finland, where coal fly ash is more expensive, this would not be the case). Taxation can impact the cost competitivity only when the mixes have a small cost price difference and a large difference in carbon footprint.

This illustration demonstrates that mixes can be categorized according to whether their price difference can be offset by taxation, and it provides a rough estimate of how large cost price differences can be compensated by taxation. However, this is a simplified calculation that does not represent a real starting point in detail. Examples of carbon prices were added to the figure only to present the magnitude of current carbon prices.

In principle, carbon emissions regulation can also have impacts, for example, on transportation costs. However, cement production or binders are the major causes of CO_2 emissions during the whole life cycle of concrete constructs; thus, the estimates of the impacts of carbon emissions regulation on the other parts of the life cycle are framed out from the scope of this study.

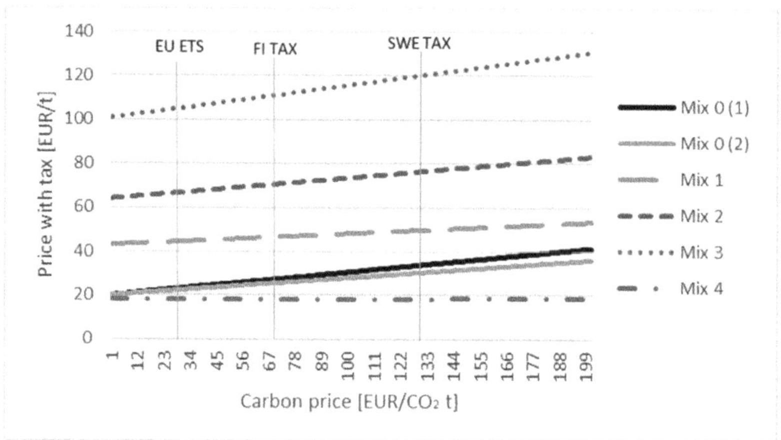

Figure 5. Cost prices of different materials with hypothetical carbon taxes [EUR/t].

3. Discussion

The cost-competitiveness of new materials is an imperative factor that determines the potential for their extensive use. So-called geopolymer concretes have been studied considerably in recent years, whereas their material costs have been limitedly studied. This study investigated how the cost-competitiveness of different materials can be assessed and what can be said about the cost-competitiveness of geopolymer concretes at a general level.

This study's approach was to conduct a preliminary cost estimate using only the "quick and dirty" approach to raw material prices alone, which could be refined, for example, by increasing transport and processing costs and refining raw material specifications. However, there are significant cost differences between some mix designs, and some mixes can be exorbitant in practice for commercial exploitation unless a clear in-service cost benefit can be presented compared to competing products.

For some geopolymer mixes, the material prices in the preliminary analysis seem to correspond to the price of traditional concrete. This requires inexpensive raw material choices, the local availability of binder precursors, and a low alkali activator proportion. This can be generalized to other recycled low-carbon concretes.

Carbon taxation cannot be expected to have a significant effect on the price differential between geopolymer concrete and conventional concrete at a general level. This is because in some geopolymer concrete mixes, the carbon footprint is similar to that of traditional concrete; the carbon footprint of OPC concrete can be expected to decrease, and for some geopolymer concrete mixes, the price difference to OPC concrete is exorbitant by default. However, taxation may change the competitive situation in rare cases, where the cost price difference of two mixes is small, but the carbon footprints differ significantly.

This study provides an overview of the cost-competitiveness of recycling-based low-carbon concretes, and the approach presented can have practical value in mix design. However, this study has some limitations. First of all, this study does not cover all costs, but focuses specifically on the effects of mix designing. On the other hand, this delineation is relevant in practice, because mix design is a determinant to the cost-competitiveness of geopolymer concretes. Secondly, with the approach described, it is not possible to accurately estimate the cost prices of material mixes for which the raw materials are not generally available for purchase. However, in these cases, the market prices of industrially produced side streams can be used as reference values for approximating what price a side stream (requiring similar processing, etc.) could be expected to have when commercially produced. Based on the target price of the material mix, target values can be set for the recycling-based raw materials that are not yet commercially produced.

4. Materials and Methods

The strategic methodological choice in this study was design science, where the aim is to develop tentative solutions to actual problems [74]. In practice, this induces case studies that can be used to answer questions about "why" or "how" [75] (p. 14).

This study was conducted as part of an EU-funded research project titled "Urban Infra Revolution," in which new geopolymer concrete mixes for urban construction were developed and piloted. Furthermore, it aimed to elucidate the market potential of these new materials. In practice, the question of market potential is crystallized into two questions: What are the benefits of using the material?; and how much does it cost? Generally, it is relevant to have a clear transparent process in making cost price estimates, especially when material development is an ongoing process that involves several parties. Therefore, this study has demonstrated a preliminary cost estimation process with illustrative examples provided by the project.

Data Collection and Analysis

In this study, the focus was on experimenting with an easy and repeatable method for material mix cost price determination. Therefore, the price search focused on Aliababa.com, the largest business to business marketplace in the world. Electronical commerce can be expected to become more widespread and centralized, and therefore the applicability of the approach can be expected to improve in the future. Market price information from other sources than this type service is laborious to collect, the information becomes rapidly obsolete, and its local utilization is limited in any case.

If the production of (geopolymer) concrete is considered in a particular location, the freights of raw materials must of course be considered in the cost estimate. However, one strategy in the geopolymer business is to import technology to geographical areas where the availability of critical side streams is good [76], so in that sense it is not a problem; even if for some materials, the price information is concentrated in certain areas (in this case, China and India). It is useful to get quickly a rough cost estimate, which can be refined, e.g., according to specific geographical areas.

Price information was sought for raw materials that commonly appear in concrete mixes presented in the literature. These data were collected from Alibaba.com, an electronic business shopping site. This service does not support downloading data directly from the system; therefore, a web scraping method was used. Web scraping, also called "parsing" or "web harvesting," means extracting data from websites. It is a form of copying selected visible data from a website, typically into a central local database or spreadsheet, for later retrieval or analysis. In this study, web scraping was conducted using Web Scraper, an extension of Google Chrome. A search was conducted using the material name as a search word. Data rows extracted included the product name, product information, including product description and price range, supplier country, and supplier name. Data were saved in comma-separated values format (csv-format) for analysis in a spreadsheet application. In this phase, the data were also filtered manually to screen out products other than searched material—for example, material processing equipment. Although this price search method has its limitations (e.g., most of the suppliers are from China), it offers easy access to extensive business to business price data. This price search was complemented with price inquiries for coal fly ash and blast furnace slag suppliers in Finland by email and phone.

Carbon dioxide emission estimates for concrete materials were collected from the scientific literature and calculated using GaBi, a database for LCA calculation. The LCA calculation was extensively described by Abdulkareem et al. [77]. Other necessary information for the calculations was collected from the literature, e.g., the current carbon prices from public sources and the material mixes from a company report.

The data were used for presenting illustrative examples of material mix cost prices. The analysis aimed to answer the questions concerning material costs that were raised by different parties during the project, such as what is the cost price of new concrete mixes compared to competing concrete mixes? How much would the carbon tax have to be to even out the price difference between OPC concrete and geopolymer concrete? How does the location of production affect the cost-competitiveness of different concretes? All of these are relevant questions when considering the commercialization of geopolymer technology.

Author Contributions: Conceptualization, A.R.; methodology, A.R.; validation, A.R. and J.H.; investigation, A.R. and M.A.; data curation, A.R. and M.A.; writing—original draft preparation, A.R. and J.H.; visualization A.R. All authors have read and agreed to the published version of the manuscript.

Funding: This research was funded by European Regional Development Fund, European Union via the Urban Innovative Actions initiative for co-financing the Urban Infra Revolution project under project number UIA02-155. The APC was funded by VITAKO ry.

Institutional Review Board Statement: Not applicable.

Informed Consent Statement: Not applicable.

Data Availability Statement: The data presented in this study are available on request from the corresponding author. The data are not publicly available due to impracticality of the data sharing services available.

Conflicts of Interest: The authors declare no conflict of interest. The funders had no role in the design of the study; in the collection, analyses, or interpretation of data; in the writing of the manuscript, or in the decision to publish the results.

References

1. Miller, S.A.; Horvath, A.; Monteiro, P.J.M. Readily implementable techniques can cut annual CO_2 emissions from the production of concrete by over 20%. *Environ. Res. Lett.* **2016**, *11*, 074029. [CrossRef]
2. Imbabi, M.S.; Carrigan, C.; McKenna, S. Trends and developments in green cement and concrete technology. *Int. J. Sustain. Built Environ.* **2012**, *1*, 194–216. [CrossRef]
3. Van Deventer, J.S.J.; Provis, J.L.; Duxson, P. Technical and commercial progress in the adoption of geopolymer cement. *Miner. Eng.* **2012**, *29*, 89–104. [CrossRef]
4. Miller, S.A.; John, V.M.; Pacca, S.A.; Horvath, A. Carbon dioxide reduction potential in the global cement industry by 2050. *Cem. Concr. Res.* **2018**, *114*, 115–124. [CrossRef]
5. Salas, D.A.; Ramirez, A.D.; Rodríguez, C.R.; Petroche, D.M.; Boero, A.J.; Duque-Rivera, J. Environmental impacts, life cycle assessment and potential improvement measures for cement production: A literature review. *J. Clean. Prod.* **2016**, *113*, 114–122. [CrossRef]
6. Cembureau. Reaching Climate Neutrality along the Cement and Concrete Value Chain by 2050—Cementing the European Green Deal. 2020. Available online: https://cembureau.eu/media/w0lbouva/cembureau-2050-roadmap_executive-summary_final-version_web.pdf (accessed on 8 July 2020).
7. Laurent, A.; Bakas, I.; Clavreul, J.; Bernstad, A.; Niero, M.; Gentil, E.; Hauschild, M.Z.; Christensen, T.H. Review of LCA studies of solid waste management systems—Part I: Lessons learned and perspectives. *Waste Manag.* **2014**, *34*, 573–588. [CrossRef]
8. *SFS-EN ISO 14040 Environmental Management—Life Cycle Assessment—Principles and Framework 2006*; Finnish Standards Association ry: Helsinki, Finland, 2006.
9. Singh, B.; Ishwarya, G.; Gupta, M.; Bhattacharyya, S.K. Geopolymer concrete: A review of some recent developments. *Constr. Build. Mater.* **2015**, *85*, 78–90. [CrossRef]
10. Hasnaoui, A.; Ghorbel, E.; Wardeh, G. Optimization approach of granulated blast furnace slag and metakaolin based geopolymer mortars. *Constr. Build. Mater.* **2019**, *198*, 10–26. [CrossRef]
11. Zhao, S.; Fan, J.; Sun, W. Utilization of iron ore tailings as fine aggregate in ultra-high performance concrete. *Constr. Build. Mater.* **2014**, *50*, 540–548. [CrossRef]
12. İlker, B.T.; Selim, Ş. Properties of concretes produced with waste concrete aggregate. *Cem. Concr. Res.* **2004**, *34*, 1307–1312. [CrossRef]
13. Yliniemi, J.; Nugteren, H.; Illikainen, M.; Tiainen, M.; Weststrate, R.; Niinimäki, J. Lightweight aggregates produced by granulation of peat-wood fly ash with alkali activator. *Int. J. Miner. Process.* **2016**, *149*, 42–49. [CrossRef]
14. Suomen Erityisjäte Oy Lehdistötiedote "Yli 300 000 Tonnia Pohjakuonaa Hyötykäyttöön". Available online: http://www.erityisjate.fi/yritys/uutiset-ja-ajankohtaiset/?newsid=309&newstitle=Lehdistötiedote+%22Yli+300+000+tonnia+pohjakuonaa+hyötykäyttöön%22 (accessed on 21 December 2018).
15. Kovářík, T.; Rieger, D.; Kadlec, J.; Křenek, T.; Kullová, L.; Pola, M.; Bělský, P.; Franče, P.; Říha, J. Thermomechanical properties of particle-reinforced geopolymer composite with various aggregate gradation of fine ceramic filler. *Constr. Build. Mater.* **2017**, *143*, 599–606. [CrossRef]
16. Aldred, J.; Day, J. Is geopolymer concrete a suitable alternative to traditional concrete. In Proceedings of the 37th Conference on Our World in Concrete, Singapore, 29–31 August 2012.

17. Albitar, M.; Mohamed Ali, M.S.; Visintin, P.; Drechsler, M. Durability evaluation of geopolymer and conventional concretes. *Constr. Build. Mater.* **2017**, *136*, 374–385. [CrossRef]

18. Valencia Saavedra, W.G.; Mejía de Gutiérrez, R. Performance of geopolymer concrete composed of fly ash after exposure to elevated temperatures. *Constr. Build. Mater.* **2017**, *154*, 229–235. [CrossRef]

19. Yang, K.-H.; Lee, K.-H.; Song, J.-K.; Gong, M.-H. Properties and sustainability of alkali-activated slag foamed concrete. *J. Clean. Prod.* **2014**, *68*, 226–233. [CrossRef]

20. Chan, C.C.S.; Thorpe, D.; Islam, M. An evaluation of life long fly ash based geopolymer cement and ordinary Portland cement costs using extended life cycle cost method in Australia. In Proceedings of the 2015 IEEE International Conference on Industrial Engineering and Engineering Management (IEEM), Singapore, 6–9 December 2015; pp. 52–56.

21. Vilamová, Š.; Piecha, M. Economic evaluation of using of geopolymer from coal fly ash in the industry. *Acta Montan. Slovaca* **2016**, *21*, 139–145.

22. Ostwal, T.; Chitawadagi, M.V. Experimental investigations on strength, durability, sustainability & economic characteristics of geo-polymer concrete blocks. *IJRET Int. J. Res. Eng. Technol.* **2014**, *3*, 115–122.

23. *SFS-EN 197-1 Cement. Part 1: Composition, Specifications and Conformity Criteria for Common Cements 2012*; Finnish Standards Association ry: Helsinki, Finland, 2012.

24. Jamieson, E.J.; Penna, B.; van Riessen, A.; Nikraz, H. The development of Bayer derived geopolymers as artificial aggregates. *Hydrometallurgy* **2017**, *170*, 74–81. [CrossRef]

25. Sun, T.; Chen, J.; Lei, X.; Zhou, C. Detoxification and immobilization of chromite ore processing residue with metakaolin-based geopolymer. *J. Environ. Chem. Eng.* **2014**, *2*, 304–309. [CrossRef]

26. Torres-Carrasco, M.; Puertas, F. Waste glass in the geopolymer preparation. Mechanical and microstructural characterisation. *J. Clean. Prod.* **2015**, *90*, 397–408. [CrossRef]

27. Yliniemi, J.; Kinnunen, P.; Karinkanta, P.; Illikainen, M.; Yliniemi, J.; Kinnunen, P.; Karinkanta, P.; Illikainen, M. Utilization of Mineral Wools as Alkali-Activated Material Precursor. *Materials* **2016**, *9*, 312. [CrossRef]

28. Bernal, S.A.; Rodríguez, E.D.; Mejia De Gutiérrez, R.; Provis, J.L.; Delvasto, S. Activation of metakaolin/slag blends using alkaline solutions based on chemically modified silica fume and rice husk ash. *Waste Biomass Valorization* **2012**, *3*, 99–108. [CrossRef]

29. Wu, Y.-H.; Huang, R.; Tsai, C.-J.; Lin, W.-T. Recycling of Sustainable Co-Firing Fly Ashes as an Alkali Activator for GGBS in Blended Cements. *Materials* **2015**, *8*, 784–798. [CrossRef]

30. Heidrich, C.; Sanjayan, J.; Berndt, M.L.; Foster, S. Pathways and barriers for acceptance and usage of geopolymer concrete in mainstream construction. In Proceedings of the 2015 World of Coal Ash (WOCA) Conference, Nasvhille, TN, USA, 5–7 May 2015.

31. RamaC References—RamaC. Available online: https://ramacreadymix.nl/en/references/ (accessed on 21 February 2020).

32. Provis, J.L.; van Deventer, J.S.J. *Alkali Activated Materials: State-of-the-Art Report, RILEM TC 224-AAM*; Springer: Dordrecht, The Netherlands, 2013.

33. van Deventer, J.S.J.; Provis, J.L.; Duxson, P.; Brice, D.G. Chemical Research and Climate Change as Drivers in the Commercial Adoption of Alkali Activated Materials. *Waste Biomass Valorization* **2010**, *1*, 145–155. [CrossRef]

34. Wagners EFC Geopolymer Pavements in Wellcamp Airport I Our Projects I Main I Wagner. Available online: https://www.wagner.com.au/main/our-projects/efc-geopolymer-pavements-in-wellcamp-airport (accessed on 5 February 2020).

35. Geocement.in. Kiran Global Green Geocement. Available online: http://www.geocement.in/product_details.php?id=25 (accessed on 21 February 2020).

36. Nu-Core Nu-Core®A2FR Non-Combustible I ACM A2FR Fire Rating, Fire Proof, Non-Combustible, Fire Retardant, Fire Rated I Products I Nu-Core. Available online: http://nu-core.com.au/nucore-fr-plus-architectural-grade-fire-rated/ (accessed on 21 February 2020).

37. Geopolymersolutions COLD FUSION CONCRETE®-FP250 I Geopolymer Solutions. Available online: https://www.geopolymertech.com/fireproofing/cold-fusion-concrete-fp250/ (accessed on 5 February 2020).

38. Sinnotec SINNOTEC—Sinnotec—Concrete Sealing und Concrete Protection—WHG-Tightened for HBV- und LAU-facilities—We Care about Concrete! Available online: http://www.sinnotec.eu/Sinnotec-Concrete-Sealing-und-Concrete-Protection-WHG-tightened-for-HBV-und-LAU-facilities/ (accessed on 21 February 2020).

39. Wagners Pinkenba Wharf I Our Projects I Main I Wagner. Available online: https://www.wagner.com.au/main/our-projects/pinkenba-wharf/ (accessed on 21 February 2020).

40. Wagners North Haven Boat Ramp I Our Projects I Main I Wagner. Available online: https://www.wagner.com.au/main/our-projects/north-haven-boat-ramp/ (accessed on 21 February 2020).

41. Schlumberger Cementing Services I Schlumberger. Available online: https://www.slb.com/drilling/drilling-fluids-and-well-cementing/well-cementing (accessed on 21 February 2020).

42. CLOCKSPRING I NRI GeoSpray®AMS Geopolymer—ClockSpring I NRI ClockSpring I NRI. Available online: https://www.clockspring.com/product/geospray-ams-geopolymer/ (accessed on 21 February 2020).

43. Omar Sore, S.; Messan, A.; Prud'homme, E.; Escadeillas, G.; Tsobnang, F. Stabilization of compressed earth blocks (CEBs) by geopolymer binder based on local materials from Burkina Faso. *Constr. Build. Mater.* **2018**, *165*, 333–345. [CrossRef]

44. Habert, G.; Ouellet-Plamondon, C. Recent update on the environmental impact of geopolymers. *RILEM Tech. Lett.* **2016**, *1*, 17. [CrossRef]

45. McLellan, B.C.; Williams, R.P.; Lay, J.; van Riessen, A.; Corder, G.D. Costs and carbon emissions for geopolymer pastes in comparison to ordinary portland cement. *J. Clean. Prod.* **2011**, *19*, 1080–1090. [CrossRef]
46. Ma, C.-K.; Awang, A.Z.; Omar, W. Structural and material performance of geopolymer concrete: A review. *Constr. Build. Mater.* **2018**, *186*, 90–102. [CrossRef]
47. Mehta, A.; Siddique, R. An overview of geopolymers derived from industrial by-products. *Constr. Build. Mater.* **2016**, *127*, 183–198. [CrossRef]
48. Reddy, M.S.; Dinakar, P.; Rao, B.H. A review of the influence of source material's oxide composition on the compressive strength of geopolymer concrete. *Microporous Mesoporous Mater.* **2016**, *234*, 12–23. [CrossRef]
49. Ng, C.; Alengaram, U.J.; Wong, L.S.; Mo, K.H.; Jumaat, M.Z.; Ramesh, S. A review on microstructural study and compressive strength of geopolymer mortar, paste and concrete. *Constr. Build. Mater.* **2018**, *186*, 550–576. [CrossRef]
50. World Bank Carbon Pricing Dashboard | Up-to-Date Overview of Carbon Pricing Initiatives. Available online: https://carbonpricingdashboard.worldbank.org/ (accessed on 4 September 2020).
51. Companies and emissions: Carbon copy. *Economist* **2013**, *322*, 70.
52. Bai, Q.; Gong, Y.; Jin, M.; Xu, X. Effects of carbon emission reduction on supply chain coordination with vendor-managed deteriorating product inventory. *Int. J. Prod. Econ.* **2019**, *208*, 83–99. [CrossRef]
53. Fareeduddin, M.; Hassan, A.; Syed, M.N.; Selim, S.Z. The impact of carbon policies on closed-loop supply chain network design. *Procedia CIRP* **2015**, *26*, 335–340. [CrossRef]
54. Jin, M.; Granda-Marulanda, N.A.; Down, I. The impact of carbon policies on supply chain design and logistics of a major retailer. *J. Clean. Prod.* **2014**, *85*, 453–461. [CrossRef]
55. Xu, J.; Chen, Y.; Bai, Q. A two-echelon sustainable supply chain coordination under cap-and-trade regulation. *J. Clean. Prod.* **2016**, *135*, 42–56. [CrossRef]
56. Zakeri, A.; Dehghanian, F.; Fahimnia, B.; Sarkis, J. Carbon pricing versus emissions trading: A supply chain planning perspective. *Int. J. Prod. Econ.* **2015**, *164*, 197–205. [CrossRef]
57. Li, D.; Wang, X.; Chan, H.K.; Manzini, R. Sustainable food supply chain management. *Int. J. Prod. Econ.* **2014**, *152*, 1–8. [CrossRef]
58. Li, J.; Su, Q.; Ma, L. Production and transportation outsourcing decisions in the supply chain under single and multiple carbon policies. *J. Clean. Prod.* **2017**, *141*, 1109–1122. [CrossRef]
59. Guo, J.; Bao, Y.; Wang, M. Steel slag in China: Treatment, recycling, and management. *Waste Manag.* **2018**, *78*, 318–330. [CrossRef]
60. Luo, Y.; Wu, Y.; Ma, S.; Zheng, S.; Zhang, Y.; Chu, P.K. Utilization of coal fly ash in China: A mini-review on challenges and future directions. *Environ. Sci. Pollut. Res.* **2021**, *28*, 18727–18740. [CrossRef] [PubMed]
61. Ganesan, N.; Abraham, R.; Deepa Raj, S. Durability characteristics of steel fibre reinforced geopolymer concrete. *Constr. Build. Mater.* **2015**, *93*, 471–476. [CrossRef]
62. Khan, M.Z.N.; Hao, Y.; Hao, H.; Shaikh, F.U.A. Mechanical properties of ambient cured high strength hybrid steel and synthetic fibers reinforced geopolymer composites. *Cem. Concr. Compos.* **2018**, *85*, 133–152. [CrossRef]
63. Al-mashhadani, M.M.; Canpolat, O.; Aygörmez, Y.; Uysal, M.; Erdem, S. Mechanical and microstructural characterization of fiber reinforced fly ash based geopolymer composites. *Constr. Build. Mater.* **2018**, *167*, 505–513. [CrossRef]
64. Vijai, K.; Kumutha, R.; Vishnuram, B.G. Effect of Inclusion of Steel Fibres on the Properties of Geopolymer Concrete Composites. *Asian J. Civ. Eng. (Build. Hous.)* **2012**, *13*, 381–389.
65. Vijai, K.; Kumutha, R.; Vishnuram, B.G. Properties of Glass Fibre Reinforced Geopolymer Concrete Composites. *Asian J. Civ. Eng. (Build. Hous.)* **2012**, *13*, 511–520.
66. Nematollahi, B.; Sanjayan, J.; Chai, J.X.H.; Lu, T.M. Properties of Fresh and Hardened Glass Fiber Reinforced Fly Ash Based Geopolymer Concrete. *Key Eng. Mater.* **2014**, *594–595*, 629–633. [CrossRef]
67. Ohno, M.; Li, V.C. An integrated design method of Engineered Geopolymer Composite. *Cem. Concr. Compos.* **2018**, *88*, 73–85. [CrossRef]
68. Behera, P.; Baheti, V.; Militky, J.; Louda, P. Elevated temperature properties of basalt microfibril filled geopolymer composites. *Constr. Build. Mater.* **2018**, *163*, 850–860. [CrossRef]
69. Li, W.; Xu, J. Mechanical properties of basalt fiber reinforced geopolymeric concrete under impact loading. *Mater. Sci. Eng. A* **2009**, *505*, 178–186. [CrossRef]
70. Nematollahi, B.; Vijay, P.; Sanjayan, J.; Nazari, A.; Xia, M.; Naidu Nerella, V.; Mechtcherine, V.; Nematollahi, B.; Vijay, P.; Sanjayan, J.; et al. Effect of Polypropylene Fibre Addition on Properties of Geopolymers Made by 3D Printing for Digital Construction. *Materials* **2018**, *11*, 2352. [CrossRef]
71. Panda, B.; Paul, S.C.; Hui, L.J.; Tay, Y.W.D.; Tan, M.J. Additive manufacturing of geopolymer for sustainable built environment. *J. Clean. Prod.* **2017**, *167*, 281–288. [CrossRef]
72. Paul, S.C.; Tay, Y.W.D.; Panda, B.; Tan, M.J. Fresh and hardened properties of 3D printable cementitious materials for building and construction. *Arch. Civ. Mech. Eng.* **2018**, *18*, 311–319. [CrossRef]
73. Abdulkareem, M.; Havukainen, J.; Horttanainen, M. How environmentally sustainable are fibre reinforced alkali-activated concretes? *J. Clean. Prod.* **2019**, *236*, 117601. [CrossRef]
74. Van Aken, J.; Chandrasekaran, A.; Halman, J. Conducting and publishing design science research: Inaugural essay of the design science department of the Journal of Operations Management. *J. Oper. Manag.* **2016**, *47–48*, 1–8. [CrossRef]

75. Yin, R.K. *Case Study Research: Design and Methods*, 5th ed.; SAGE: Thousand Oaks, CA, USA, 2014.
76. Karttunen, E.; Tsytsyna, E.; Lintukangas, K.; Rintala, A.; Abdulkareem, M.; Havukainen, J.; Nuortila-Jokinen, J. Toward environmental innovation in the cement industry: A multiple-case study of incumbents and new entrants. *J. Clean. Prod.* **2021**, *314*, 127981. [CrossRef]
77. Abdulkareem, M.; Havukainen, J.; Nuortila-Jokinen, J.; Horttanainen, M. Life cycle assessment of a low-height noise barrier for railway traffic noise. *J. Clean. Prod.* **2021**. submitted.

recycling

MDPI

Article

Mechanical Properties of Tin Slag Mortar

Nathaniel Olukotun [1,2,*], Abdul Rahman Mohd Sam [1,3], Nor Hassana Abdul Shukor Lim [1], Muyideen Abdulkareem [4], Isa Mallum [1,5] and Olukotun Adebisi [6]

[1] School of Civil Engineering, Universiti Teknologi Malaysia, Johor 81310, Malaysia; abdrahman@utm.my (A.R.M.S.); norhasanah@utm.my (N.H.A.S.L.); misa@graduate.utm.my (I.M.)
[2] Department of Building Technology, Kogi State Polytechnic, Lokoja 270103, Nigeria
[3] Construction Research Centre, Universiti Teknologi Malaysia, Johor 81310, Malaysia
[4] Department of Civil Engineering, Faculty of Engineering, Technology and Built Environment, UCSI University, Kuala Lumpur 56000, Malaysia; abdulkareem@ucsiuniversity.edu.my
[5] Department of Civil Engineering Technology, Adamawa State Polytechnic, Yola 640231, Nigeria
[6] Department of Civil Engineering, University of Abuja, Abuja 900108, Nigeria; olukotunadebisi@gmail.com
* Correspondence: olukotunnathaniel78@gmail.com

Abstract: The increased demand for cement mortar due to rapid infrastructural growth and development has led to an alarming depletion of fine aggregate. This has prompted the need for a more sustainable material as a total/partial replacement for natural fine aggregate. This study proposes the use of tin slag (TS) as a replacement for fine aggregate in concrete to bridge this sustainability gap. TS was used to replace fine aggregate at replacement levels of 0%, 25%, 50%, 75%, and 100% in cement mortar. Fresh and hardened properties of TS mortar were obtained. Flow tests showed that, as the TS quantity and the w/c ratio increased, the mortar flow increased. Similarly, the compressive strength increased as the TS replacement increased up to 50% replacement, after which a decline in strength was observed. However, with the TS replacement of fine aggregate up to 100%, a compressive strength of 6% above control was attained. The morphological features confirm that specimens with TS had a denser microstructure because of its shape characteristics (elongated, irregular, and rough), and, thus, plugged holes better than the control mortar. The natural sand's contribution to strength was a result of better aggregate hardness as compared to TS. Hence, TS can be used as alternative for fine aggregate in sustainable construction.

Keywords: tin slag; mortar; compressive strength; fine aggregate; rough surfaced; elongated

Citation: Olukotun, N.; Sam, A.R.M.; Lim, N.H.A.S.; Abdulkareem, M.; Mallum, I.; Adebisi, O. Mechanical Properties of Tin Slag Mortar. *Recycling* **2021**, *6*, 42. https://doi.org/10.3390/recycling6020042

Academic Editors: José Neves and Ana Cristina Freire

Received: 15 January 2021
Accepted: 31 March 2021
Published: 21 June 2021

Publisher's Note: MDPI stays neutral with regard to jurisdictional claims in published maps and institutional affiliations.

1. Introduction

Concrete is the most utilized material after water. Consequently, its demand has been on the rise at a rate of 5% annually. It is estimated that about 25 to 30 billion tons of concrete are manufactured globally every year [1]. This amount of produced concrete suggests that its constituents are exploited in huge volumes. This may further increase because of population explosion, unprecedented growth, and global industrialization. Fine aggregate constitutes about 30% of the volume of concrete. There are no exact measures on the quantity of fine aggregates available for use worldwide or by country. However, from 2018 estimates of cement consumed worldwide, about 8 billion tons of fine aggregate is consumed yearly [1]. It has been reported that the rate of depletion is faster than its renewal, and the consumption rate increases every year. In addition, natural sand is utilized for the production of glass and electronics and for road construction.

Reducing the consumption of natural sand by looking for alternative materials for its replacement is important. Lately, crushed and manufactured sand have been used as alternatives to natural fine aggregate (NFA) [2,3]. However, the crushing and manufacturing process is capital intensive. Many studies have been conducted to establish new substitutes for NFA by utilizing industrial waste materials such as steel slag [4], lead slag [5], copper slag [6], foundry sand [7], and recycled glass [8]. This reduces the dependence on natural sand and propels the sustainability of this declining resource. For example, Zhao et al. (2018) replaced 20% of fine aggregate in concrete with steel slag. The replacement showed that compressive strength was improved by 28% over the control concrete sample. Similarly, copper slag was used up to 100% as a fine aggregate replacement by Prem et al. (2018) [6]. It was observed that 100% copper mortars had a 3% compressive strength above the control mortar replacement. The use of foundry sand and recycled glass at 10% replacement has also been found to improve the mechanical properties of concrete [9,10].

Another such industrial waste readily available and under-utilized is tin slag (TS). TS is a by-product obtained from the production of tin. About 2 million tons of TS waste is available at landfills worldwide [11]. Very few studies have been conducted on the use of TS as a fine aggregate replacement. Shakil and Hassan [12] evaluated the properties of TS polyester polymer concrete. The compressive strength test showed that samples attained a strength of 125.07 MPa. Hashim et al. [13] applied TS as a replacement of fine aggregate to produce interlocking pavement bricks. The study showed that the compressive strength increased by 20%, however, the study considered only 100% and 20% TS replacement of fine aggregate. Rustandi et al. (2018) [14] utilized the pozzolanic properties of TS by replacing 10% of the cement. The TS mortar achieved only a 63% compressive strength of the control sample at 28 days.

This study examines the utilization of TS as a replacement of NFA in cement mortar. Five different replacement fractions of TS (0%, 25%, 50%, 75%, and 100%) were introduced into the mortar mixes as replacements of fine aggregate. This was done using different water/cement (w/c) ratios of 0.5, 0.55, and 0.6. The specimens were prepared, and tests were conducted at fresh and hardened states. Test were conducted after 3, 7, and 28 days of water-curing the specimens. The workability of the mortar was measured by the flow table test to evaluate the effect of TS on the flowability of TS mortars. The mechanical properties measured were compressive strength, flexural strength and tensile strength

2. Experiment

2.1. Materials

The materials used in this study were cement, NFA, TS, and water. TS was used as a replacement material of fine aggregate in different weight fractions. Ordinary Portland cement (OPC) with a gauge of 42.5 and a specific gravity of 3.14 according to ASTM C (American Standard Testing Methods) 150 [15] was used. NFA and TS of a 4.75 mm maximum aggregate size were used. The TS used was obtained from the tin smelting company in Penang, Malaysia. Tap water was used for the mixing and curing of the mortar specimens. The physical and chemical properties of the fine aggregate were measured according to ASTM C128 [16] and presented in Tables 1 and 2. The shape characteristics showed that the fine aggregate particles were round and smooth, whereas the TS particles were elongated and rough-surfaced. TS particles at the macro state are depicted in Figure 1.

2.2. Particle Size Gradation

The grading of aggregates was done in conformity to ASTM C33 [17] protocols. The initial sieve analysis performed for TS aggregate indicated that it did not conform to ASTM requirements, as shown in Figure 2. About 50% of the aggregate fell within the 4.75 to 2.36 range. The grading method used was to separate the aggregate passing through different sieves and thereafter combine the appropriate percentage allowed of each size according to ASTM standards. The aggregates were thoroughly mixed and a second sieve analysis was conducted. Figure 3 shows the chart of the second sieve analysis. Large aggregates (4.75–2.36 mm) that were in excess were grinded in a ball mill machine to reduce their sizes and were thereafter re-graded. This was necessary to ensure that aggregates were efficiently utilized and that wastage avoided. About 2% of aggregates were discarded.

Table 1. Physical properties of fine aggregate and tin slag (TS).

S/N	Physical Properties	Sand	TS
1	Specific gravity	2.62	3.0
2	Water absorption	2.0	1.98

Table 2. Chemical properties showing major oxides of fine aggregate and tin slag.

S/N	Chemical Properties	Sand	TS
1	Silica	91.5	15.8
2	Alumina	3.55	7.10
3	Magnesium oxide	1.3	0.85
4	Iron oxide	1.2	33.4
5	Calcium oxide	0.5	21.3
6	Titanium(IV) oxide	-	5.74
7	Zirconium oxide	-	2.16
9	Niobium(V) oxide	-	1.36
10	Tin(IV) oxide	-	1.62
11	Tantalum(V) oxide	-	1.31
12	Tungsten(VI) oxide	-	3.08

Figure 1. The macro morphology of tin slag.

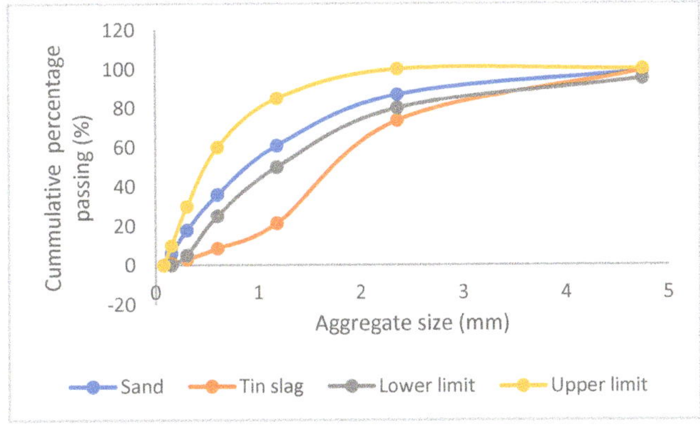

Figure 2. Initial sieve analysis.

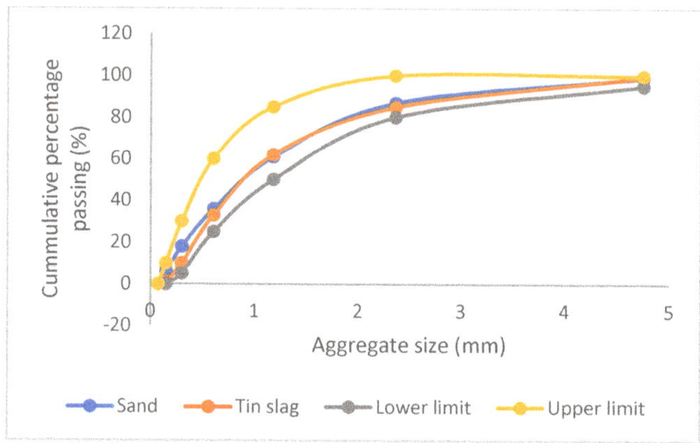

Figure 3. Second sieve analysis.

2.3. Mix Proportioning and Sample Preparation

Fifteen different mixes were prepared for this study. All mortar specimens were prepared using a cement to fine aggregate volume ratio of 1:3. The control mix sample was NFA-based, while the other samples had 25%, 50%, 75%, or 100% of the fine aggregate replaced with TS. The TS percentage replacement levels of fine aggregate are shown in Table 3. The control sample (TS0) contained a 0% TS replacement of fine aggregate, whereas TS100 contained a 100% TS replacement of fine aggregate. The w/c ratios applied in this study were 0.5, 0.55, and 0.6.

Table 3. The mix percentages of tin slag (TS) in mortar.

Designation	Fine Aggregate (%)	Replacement Levels of TS (%)
Control (TS0)	100	0
TS25	75	25
TS50	50	50
TS75	25	75
TS100	0	100

It should be noted that the TS used in this study had a specific gravity of 14.5% above that of NFA. Therefore, multiplication factors (shown in Table 4) were applied to obtain the correct corresponding TS mass percentage replacement of fine aggregate. These multiplication factors were applied so that aggregate volumes of different mortar mixes (TS0, TS25, TS50, TS75, and TS100) were the same, despite the difference in the specific gravities of the NFA and the TS. The multiplication factors were derived from the densities of the aggregates (natural sand and TS) used for the different mixes. For example, the multiplication factor for TS25 = the density of the combined aggregates (75% NFA and 25% TS)/the density of the NFA. The constituent quantity of each mix sample is given in Table 5. To reduce error in this study, three specimens were produced for each mix proportion, and the average was taken. In this study, 135 cubes were prepared for compressive strength tests, while 45 cylinders and 45 prisms were prepared for splitting tensile and flexural strength tests, respectively. In addition, 45 mortar bars were prepared for ASR (Alkali silica reaction) tests. Demolding was done after 24 h and samples were cured in water for 3-, 7-, and 28-day periods.

Table 4. The Multiplier factors for tin slag mortars.

Designation	Multiplier Factor
TS25	1.06
TS50	1.12
TS75	1.16
TS100	1.20

Table 5. The mix proportions of constituents for 1 m^3 of mortar.

Mix	Designation	Cement (kg)	Sand (kg)	Tin Slag (kg)	Water (kg)	*w/c*
M1	TS0	530	1590	0	265	0.5
M2	TS25	530	1192.5	421	265	0.5
M3	TS50	530	795	890	265	0.5
M4	TS75	530	397.5	1383	265	0.5
M5	TS100	530	0	1908	265	0.5
M6	TS0	530	1590	0	291.5	0.55
M7	TS25	530	1192.5	421	291.5	0.55
M8	TS50	530	795	890	291.5	0.55
M9	TS75	530	397.5	1383	291.5	0.55
M10	TS100	530	0	1908	291.5	0.55
M11	TS0	530	1590	0	318	0.6
M12	TS25	530	1192.5	421	318	0.6
M13	TS50	530	795	890	318	0.6
M14	TS75	530	397.5	1383	318	0.6
M15	TS100	530	0	1908	318	0.6

2.4. Testing Procedures

Several tests were carried out on the TS cement mortar and control samples to evaluate the effects of the replacement of fine aggregate with TS. These involved the fresh and hardened states of the samples. The fresh property considered in this study was workability, and this was performed through a mortar flow test in accordance to ASTM C 1437 [18] guidelines.

The hardened properties measured were compressive strength, tensile strength, flexural strength, and alkali–silica reactivity. The compressive strength was measured using a 3000 kN compression testing machine in accordance to ASTM C109 [19] procedures. Cube dimensions of 50 × 50 × 50 mm^3 were used in this experiment. The tensile strength test was performed according to ASTM C 496 [20] guidelines. Samples used for tensile strength tests were 70 mm (diameter) × 150 mm (length) cylinders. Testing was done with plywood strips positioned perpendicular to diametral lines drawn on the specimens.

The flexural strength testing, on the other hand, was performed according to ASTM 348 [21]. The specimens applied for the flexural test had dimensions of 40 × 40 × 160 mm^3. Figure 4 shows the hardened state tests (compressive, tensile, and flexural tests). Additionally, the alkali–silica reactivity check was performed according to ASTM C1567. This test was performed to determine the level of reactivity of the TS aggregate in the mortar. This was necessary because of the presence of CaO in TS, as shown in Table 2. The phase diagram of grounded TS (Figure 5) shows that it has a dominant amorphous phase with few crystalline peaks. The specimens used were 25 × 25 × 250 mm^3 mortar bars.

(a)

(b)

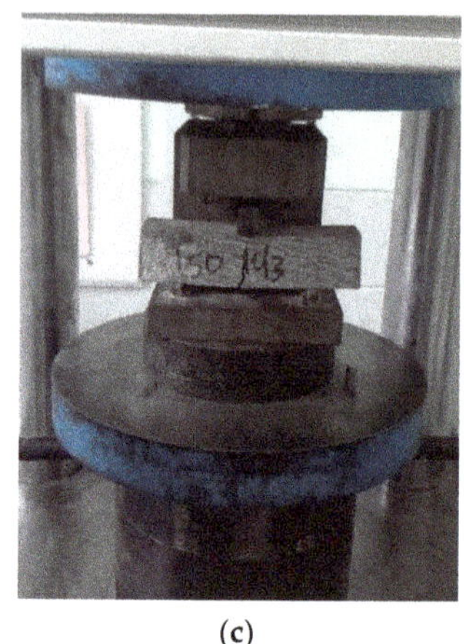
(c)

Figure 4. Hardened state tests: (**a**) the compression test, (**b**) the split tensile test, and (**c**) the flexure test.

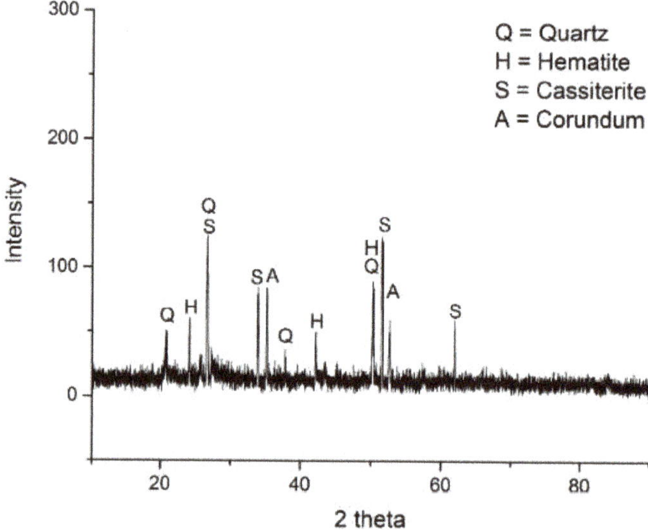

Figure 5. XRD (X-ray diffractometer) pattern of tin slag aggregate [14].

3. Experimental Results

3.1. Workability

Workability is the property of mortar that allows for adequate compaction, placement, and finish without segregation and/or bleeding. The flowability for all samples containing graded and non-graded particles were measured by the flow cone test. Figure 6 illustrates the mortar flow after the flow cone tests. As can be seen, normal and scattered flows were observed. All samples of graded and ungraded TS mortars exhibited normal flow, except ungraded TS100 mortars of w/c ratios of 0.5 and 0.55. The flow test results for the ungraded and graded particles are presented in Figures 7 and 8.

(a) **(b)**

Figure 6. Mortar consistencies: (**a**) normal flow and (**b**) scattered flow.

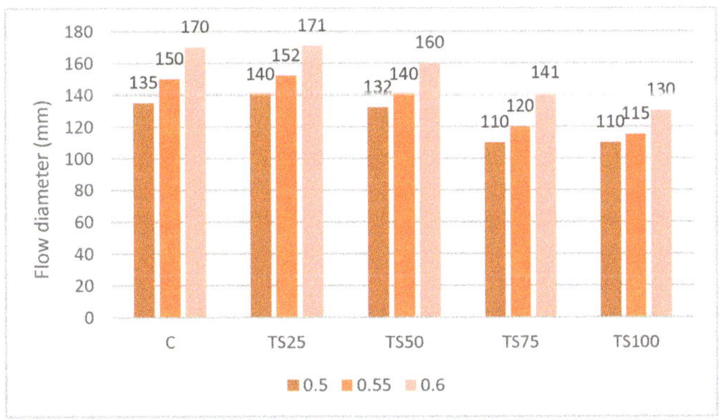

Figure 7. Flow diameters of ungraded TS at different *w/c* ratios.

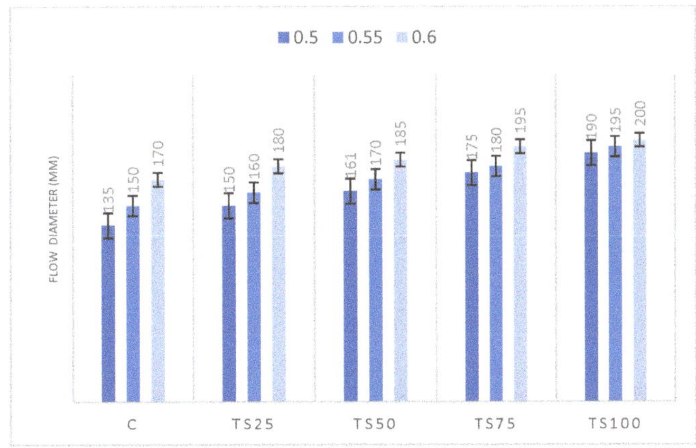

Figure 8. Flow diameters of graded TS at different *w/c* ratios.

Figure 7 indicates that the workability of ungraded TS mortar marginally increased up to a 25% replacement, but, thereafter, a downward trend were observed for TS50, TS75, and TS100 mortars.

The workability of the control sample and of TS25 was 170 mm and 171 mm, respectively, at a w/c ratio of 0.6. As the percentage of ungraded TS increased, the workability of the TS mortar decreased, provided that the w/c ratio remained the same. For example, at a w/c ratio of 0.6, the workability increased from 170 mm to 171 mm for TS0 and TS25, and afterwards decreased to 160 mm, 141 mm, and 130 mm for TS50, TS75, and TS100 mortars, respectively. This effect is attributed to particle interference, which increasingly occurred as TS replacement increased. The larger particles (2.36–4.5 mm) were more prominent smaller particles (Figure 2). Thus, pockets of spaces narrower than the diameter of the smaller particles were created. These spaces trapped water that would have provided better consistency, thereby increasing the water demand [22]. In addition, due to the lack of smaller particles in the ungraded TS, the ball bearing effect by smaller particles (which is needed to lubricate larger particles) was reduced as TS replacement increased [23]. This trend was noticed at w/c ratios of 0.5 and 0.55, except regarding the TS 100 mortar, for which a scattered flow was observed. The mortar flow diameter does not give an accurate representation of flow. The scattered flow was attributed to inadequate cohesion between TS aggregates within the matrix, which is due to insufficient water for mixing [24]. Pockets of spaces created by the ungraded TS used up part of the mixing water. Figure 8 shows that the workability of graded TS mortar increased as the quantity of TS increased [25,26]. For example, the workability of graded TS mortar increased from 135 to 190, as the quantity of TS increased from 0 to 100% at a w/c ratio of 0.5. Cumulative increases of 10 mm, 15 mm, 30 mm, and 35 mm above the control were noticed for TS25, TS50, TS75, and TS100 mortars. The same trend was noticed at a w/c ratios of 0.55 and 0.6.

From Figures 7 and 8, it can be observed that higher w/c ratio yielded increased workability, both for graded and ungraded TS samples. For example, when the w/c ratio of the ungraded TS was increased from 0.5 to 0.6 at TS50, the workability increased from 132 mm to 160 mm, while an increase from 161 mm to 185 mm was observed for graded TS samples. This was due to an increase in water for the mixing constituents as the water content increased. However, in terms of shape and texture, round and smooth particles have better workability than elongated or angular particles [23,27]. Nonetheless, it was observed that graded TS mortars had higher flow diameters than ungraded TS mortars. This was attributed to the low water absorption of TS, which promoted water retention which improved lubrication between particles. Thus, the effect of yield stress and water demand resulting from the angular shape characteristics of TS was insignificant [28].

3.2. Alkali–Silica Reactivity

Figure 9 shows the result of the alkali–silica reaction test of samples in this study. The expansion of TS mortars at different replacement levels are presented. The drying shrinkage of TS100 mortars were just fractionally lower than the control. Thus, the differential effects on these mortars are insignificant. The shrinkage values of mortars at 14 days for TS25, TS50, TS75, and TS100 were 0.009%, 0.008%, 0.0096%, and 0.0097%, respectively, compared to the control specimen (TS0) value of 0.01%. The T50 mortar had the lowest expansion of 0.08%. The lower rate of expansion of TS mortars was attributed to the rough texture and angularity of aggregates, which provided a denser microstructure. This has been reported by other researchers [29,30]. However, the contribution of the fine aggregate's hardness contributed to the compactness observed in TS50 mortars. Overall, all mortars had comparable results and were within the ASTM [31] expansion limit of 0.1%; therefore, TS aggregate is non-reactive.

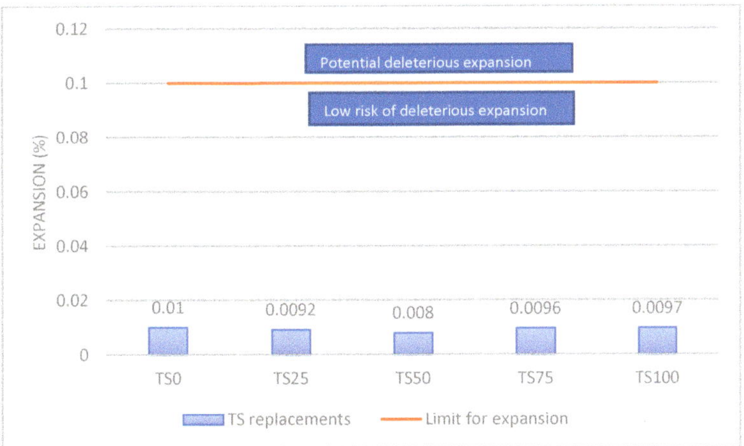

Figure 9. The expansion of the mortars at 14 days.

3.3. Compressive Strength

In this section, the 28-day compressive strength of graded TS mortars with different w/c ratios are presented and compared to the control sample. In addition, the compressive strengths of the TS cement mortar and control samples obtained at days 3, 7, and 28 are presented to check the early strength of the TS cement mortar samples. Figure 10 shows the 28-day compressive strength of mortar mixes at different w/c ratios. In all samples (control and TS), It was observed that, as w/c increased, the compressive strength decreased. The compressive strength decreased by as much as 25% due to the increased w/c ratio. For example, at TS50, the compressive strength decreased from 45.42 MPa to 34.22 MPa when the w/c ratio increased from 0.5 to 0.6. This result is consistent with previous findings [32,33]. However, to achieve good workability, mortars must not exhibit segregation or bleeding. Samples with a w/c ratio of 0.5 had the highest compressive strength but were too stiff; thus, their use is not practicable. Additionally, mortars with a w/c ratio of 0.6 were most workable, but segregation and bleeding were noticed, as was low compressive strengths (Figure 6). Contrarily, bleeding and segregation did not occur in mixes with a w/c ratio of 0.55 Thus, mortar mixes containing a water to binder ratio of 0.55 were selected based on their workability and strength. Compressive strength development is presented in Figure 11. However, both ungraded and graded TS mortar samples were tested for compressive strength to evaluate the influence of grading, as workability results showed differing trends (as shown in Figures 7 and 8).

Figure 11 shows the compressive strength development of mortar samples with a w/c ratio of 0.55 after 28-days of curing. All fine aggregate replacements with graded TS increased the compressive strength of cement mortar on all curing days (3, 7, and 28). For example, the 3, 7, and 28 days compressive strengths of the control sample were 25.43 MPa, 33.87 MPa, and 38.75 MPa, respectively, whereas the TS samples' lowest values were 26.35 MPa, 34.5 MPa, and 40.2 MPa, respectively. The increase in strength was due to the improved bonding provided by the TS's aggregate angularity and surface roughness. Furthermore, as TS increased in the cement mortar, the compressive strength also increased. This compressive strength increase peaked at 50% replacement (TS50), while further increase in TS quantity decreased the compressive strength of the TS mortar. For example, at 28 days, when TS increased from 25% (TS25) to 50% (TS50), the compressive strength increased from 40.2 MPa to 43.15 MPa. Thereafter, this value decreased to 42.99 MPa and 41.12 MPa as TS increased to 75% (TS75) and 100% (TS100), respectively. The decrease in compressive strength is attributed to the reduction of aggregate hardness as TS% increased to 75% and 100% [34].

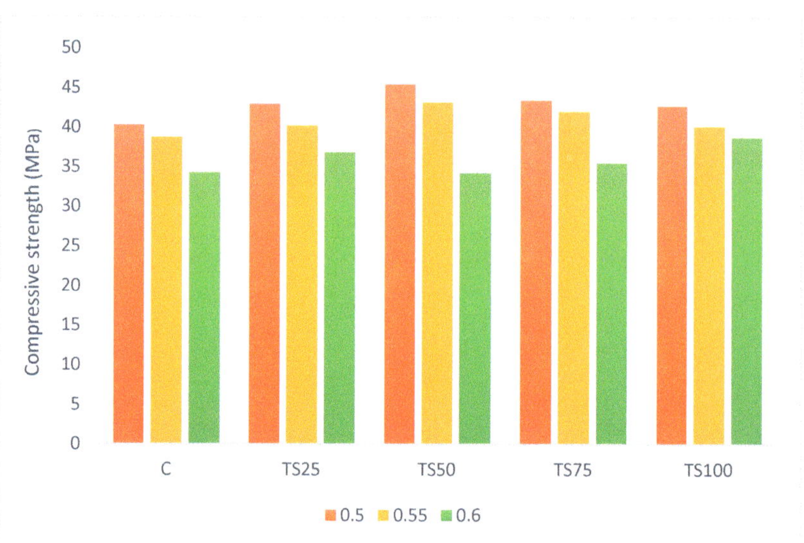

Figure 10. The compressive strength of graded TS at different w/c ratios.

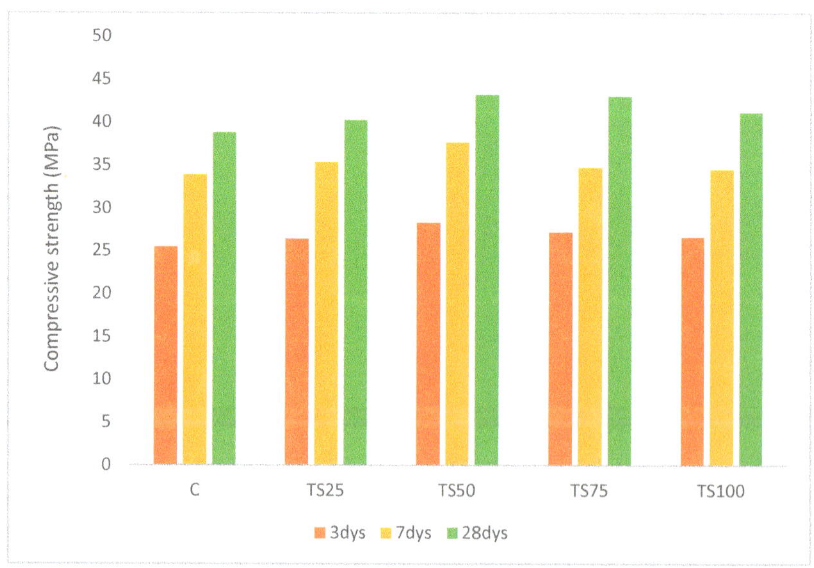

Figure 11. The compressive strength development of graded TS mortars at a w/c of 0.55.

Compressive strength was improved because of the densification of TS mortars, which is linked to TS's rough surface structure and the well-bonded interfacial transition zone (ITZ) of TS mortars. Better adherence was expected for rough-surfaced materials compared to smooth-surfaced fine aggregate. The elongated and angular shape of the TS aggregate also contributed to the filling pores between the TS mortar a greater extent than the control mortars. The elongation of TS aggregates allowed a greater depth of embedment while their angularity improved the bond within the matrix. Similar results were reported by Ouda and Abdel-Gawwad [35] and Guo et al. [36]. Ouda and Abdel-Gawwad observed that the rough surface and angularity of oxygen furnace slag aggregate are attributed to the progressive contribution of strength up to 100% replacement. Likewise, Guo et al. indicated that the rough surface and angularity of steel slag resulted in a higher bonding strength. On the other hand, the contribution of natural sand was attributed to the hardness of the natural sand aggregate as compared to the TS. Thus, the combined effect of the fine aggregate strength of natural sand and the good adherence and bonding characteristics of the TS aggregate provided the optimum strength recorded at 50% replacement. Conversely, when the TS content increased above 50%, the contribution of natural sand to compressive strength in terms of hardness reduced. This is because the contribution of natural aggregate reduces while that of TS increases as the amount of TS increases beyond 50%. However, the effects of aggregate hardness and bonding characteristics were optimal at 50% replacement. Thus, the optimum strength was obtained at 50% replacement.

A different trend was observed with the ungraded TS (Figure 12). A progressive reduction in the compressive strength was noticed, even at 50% replacement, which had the highest compressive strength for graded TS.

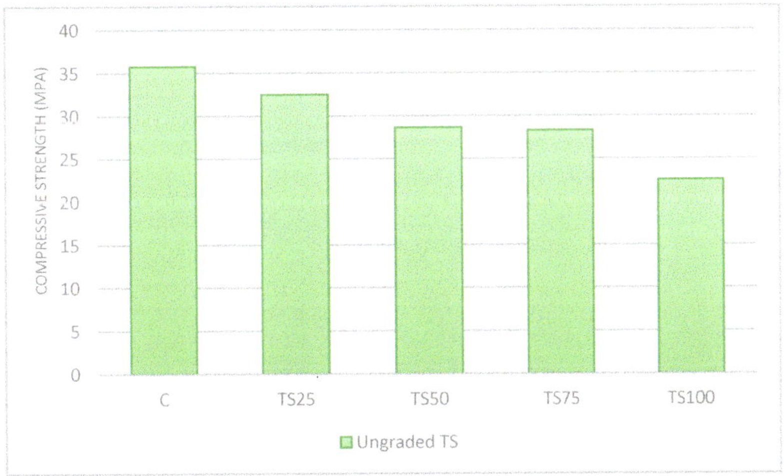

Figure 12. The 28-day compressive strength of ungraded TS mortars at a w/c of 0.55.

The control mortar had the highest compressive strength of 35.78 MPa, whereas TS25, TS50, TS75, and TS100 had compressive strengths of 32.51 MPa, 28.67 MPa, 28.33 MPa, and 22.54 MPa, respectively. This shows that the effect of the TS aggregate in terms of bond improvement in graded TS was not seen in the ungraded TS. This was due to inadequate distribution of aggregates within the matrix because larger aggregates dominate as TS replacement increases. Consequently, more pores are formed within the matrix and the progressive compressive strength reduces.

Figure 13 shows the strength development in graded and ungraded samples. An upward trend was observed with graded TS and a downward trend for ungraded TS samples. The increase in strength noticed in the graded TS mortar was attributed to improved packing density, as smaller aggregates fill pore spaces between larger aggregates. This was affirmed in Li et al. [37]. It was observed that packing density is a major factor that influences the compressive strength behavior of cement composites. The compactness of composites produced from mixed aggregates slows down crack propagation because of improved packing density as pore spaces are filled by smaller aggregates. This results in a smaller number of cracks and crack diameters, as compared to uniform or ungraded aggregates. On the other hand, in ungraded samples, a progressive reduction in compressive strength was noticed as TS replacement increased. This was credited to the progressive increase in particle interference due to the increasing number of larger particles. Thus, packing density was not optimized, so a larger number of pores and, consequently, a low compressive strength resulted.

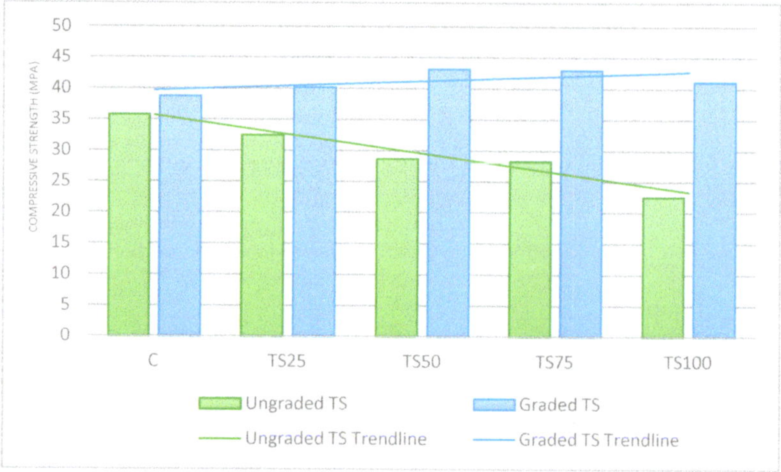

Figure 13. 28-day compressive strength of ungraded and graded TS mortars at a w/c of 0.55.

3.4. Splitting Tensile Strength

Cement composites such as mortar are generally weak while in tension. Under tension, the ITZ bears the tensile stresses for the matrix to be together. The strength of the ITZ is much weaker than that of the aggregates; thus, disintegration occurs as tensile stresses increases beyond the ITZ strength. Figure 14 shows the tensile strength of graded TS mortar samples with a w/c ratio of 0.55 at 3, 7, and 28 days. The tensile strength increased as number of curing days increased. For example, for TS25, when number of curing days increased from 3 days to 28 days, the tensile strength increased from 3.08 MPa to 4.1 MPa. The tensile strengths obtained at 28 days were similar for all the TS and control samples. The control sample C, TS25, and TS50 achieved tensile strengths of 4.08 MPa, 4.1 MPa, and 4.15 MPa, respectively. Similar findings were observed after 7 days of curing in all considered samples. However, after 3 days of curing, the tensile strength increased for TS25 and TS50, but decreased to values close to the control sample C. The 3-day tensile strength increased from 2.63 MPa for TS0 up to 3.08 MPa for TS50, and afterwards there was a decline to 2.62 and 2.57 for the TS75 and TS 100 mortars, respectively. The higher tensile strength observed in the TS50 mix was due to the greater contribution of the hardness effect by natural sand and the bonding effect in TS when compared to other mixes. However, Waheed [38] observed a significant 35% increase in concrete strength over the control, when 20% of TS was used for sand replacement. Therefore, the gains in strength in relation to aggregates might slightly differ from mortar because of the involvement of coarse aggregates in concrete.

3.5. Flexural Strength

Flexural strength is sensitive to the aggregate type and little variations in specimen preparation and testing. Figure 15 shows the results of the flexural strength test of the graded TS mortar samples. The test was conducted on the TS and control samples after 3, 7, and 28 days of water curing. It was observed that, as the curing days increased, the flexural strength increased for the control and for all TS samples. Flexural strength gain of 11% to 15% was recorded from 7 to 28 days curing periods across all replacement levels. In addition, it was observed that the flexural strength increased as TS levels increased up to 75%, but peaked at 50%. However, a further increase in TS (TS100) resulted in decreased flexural strength. At 28 days of curing, the flexural strengths of TS25, TS50, TS75 and TS100 were 12.00 MPa, 12.35 MPa, 11.79 MPa and 11.48 respectively, whereas the control sample was 11.54 MPa. This shows that compressive strength at all replacement levels up to 100% were comparable to the control. The bond properties of the TS mortar matrix are influenced by the irregular and elongated aggregate shape.

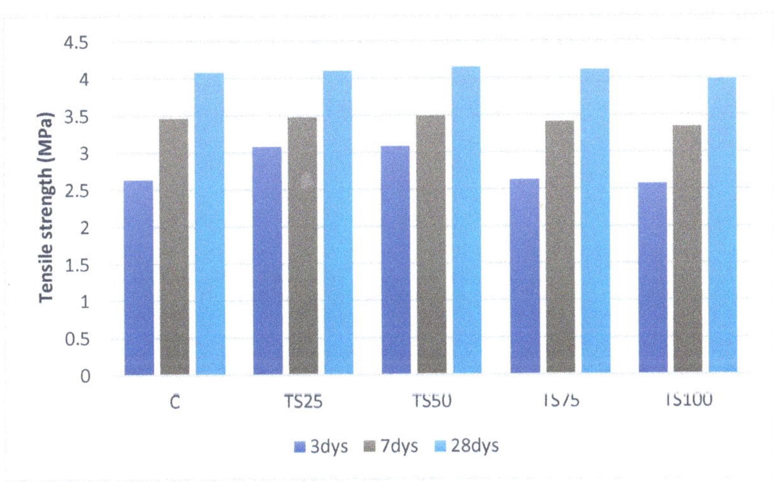

Figure 14. The tensile strength across curing days at a w/c of 0.55.

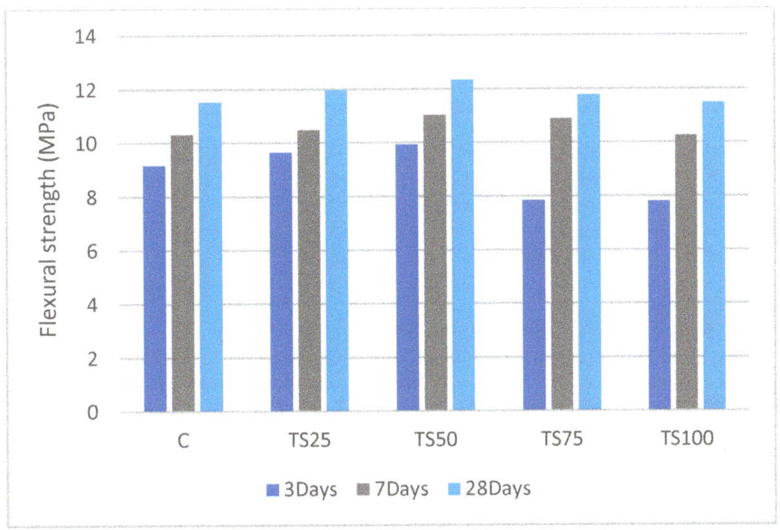

Figure 15. Flexural strength of mortars with a w/c of 0.55 at 3, 7, and 28 days of curing.

The angularity and distribution of aggregates in the cement paste played a significant role in the process of load transfer [39]. In normal-strength concrete, cracks are mostly propagated around the aggregate particles [40]. Thus, a stronger ITZ and aggregate particle interlocking in TS mortar reduces the impact on the failure plane as the stresses induced are distributed.

4. Conclusions

This study was undertaken in order to evaluate the performance of TS mortar at different TS replacements. The findings show that TS aggregate can be used to replace NFA. Based on the findings of this experimental research, the following conclusions are drawn:

- Grading has a great effect on the fresh properties of TS mortar. Ungraded mortar was observed to have reduced workability as TS replacement increased, whereas grading improved the workability and reduced the water demand of graded TS mortar.
- Mortar containing 50% showed higher compressive strength than the reference mortar at all ages. Additionally, TS100 mortars had a compressive strength comparable to the control.
- The splitting tensile and flexural strengths of tin slag mortars were higher than the corresponding control mortars.
- The expansions of TS mortars were below allowable limits of 0.1%.
- TS could be used as a substitute for natural sand for the production of mortars. This would minimize cost and natural resource depletion, thus engendering the sustainability of fine aggregate.

Author Contributions: N.O.: Investigation and Writing—original draft; A.R.M.S.: Supervision, review/editing and funding acquisition; N.H.A.S.L.: Supervision and funding acquisition; M.A.: Writing—review & editing; I.M.: Writing—review & editing; O.A.: Writing—review & editing. All authors have read and agreed to the published version of the manuscript.

Funding: This research was funded by CRG grant: R.J130000.7309.4B554.

Data Availability Statement: Data can be made available from the authors.

Conflicts of Interest: The authors declare no conflict of interest.

References

1. Klee, H. The cement sustainability initiative. *Cem. Sustain. Initiat.* **2017**, *157*, 9–12.
2. Li, H.; Huang, F.; Cheng, G.; Xie, Y.; Tan, Y.; Li, L.; Yi, Z. Effect of granite dust on mechanical and some durability properties of manufactured sand concrete. *Constr. Build. Mater.* **2016**, *109*, 41–46. [CrossRef]
3. Makhloufi, Z.; Kadri, E.H.; Bouhicha, M.; Benaissa, A.; Bennacer, R. The strength of limestone mortars with quaternary binders: Leaching effect by demineralized water. *Constr. Build. Mater.* **2012**, *36*, 171–181. [CrossRef]
4. Zhao, Z.; Qu, X.; Li, F.; Wei, J. Effects of steel slag and silica fume additions on compressive strength and thermal properties of lime-fly ash pastes. *Constr. Build. Mater.* **2018**, *183*, 439–450. [CrossRef]
5. Ogundiran, M.B.; Nugteren, H.W.; Witkamp, G.J. Immobilisation of lead smelting slag within spent aluminate-fly ash based geopolymers. *J. Hazard. Mater.* **2013**, *248–249*, 29–36. [CrossRef]
6. Prem, P.R.; Verma, M.; Ambily, P.S. Sustainable cleaner production of concrete with high volume copper slag. *J. Clean. Prod.* **2018**, *193*, 43–58. [CrossRef]
7. Prasad, V.D.; Prakash, E.L.; Abishek, M.; Dev, K.U.; Kiran, C.K.S. Study on concrete containing Waste Foundry Sand, Fly Ash and Polypropylene fibre using Taguchi Method. *Mater. Today Proc.* **2018**, *5*, 23964–23973. [CrossRef]
8. Tamanna, N.; Tuladhar, R.; Sivakugan, N. Performance of recycled waste glass sand as partial replacement of sand in concrete. *Constr. Build. Mater.* **2020**, *239*. [CrossRef]
9. Pandey, P.; Harison, A.; Srivastava, V. Utilization of Waste Foundry Sand as Partial Replacement of Fine Aggregate for Low Cost Concrete. *Inpressco* **2015**, *5*, 3535–3538.
10. Kashani, A.; Ngo, T.D.; Hajimohammadi, A. Effect of recycled glass fines on mechanical and durability properties of concrete foam in comparison with traditional cementitious fines. *Cem. Concr. Compos.* **2019**, *99*, 120–129. [CrossRef]
11. Izard, C.F.; Müller, D.B. Tracking the devil's metal: Historical global and contemporary U.S. tin cycles. *Resour. Conserv. Recycl.* **2010**, *54*, 1436–1441. [CrossRef]
12. Shakil, U.A.; Bin, S.; Hassan, A. Behavior and properties of tin slag polyester polymer concrete confined with FRP composites under compression. *J. Mech. Behav. Mater.* **2020**, *29*, 44–56. [CrossRef]
13. Hashim, M.J.; Mansor, I.; Ismail, M.P.; Sani, S. Preliminary study of tin slag concrete mixture. In Proceedings of the IOP Conference Series: Materials Science and Engineering, International Nuclear Science, Technology and Engineering Conference 2017 (iNuSTEC2017), Selangor, Malaysia, 25–27 September 2017; Volume 298. [CrossRef]
14. Rustandi, A.; Nawawi, F.; Cahyadi, A. Evaluation of the suitability of tin slag in cementitious materials: Mechanical properties and Leaching behaviour Evaluation of the suitability of tin slag in cementitious materials: Mechanical properties and Leaching

behaviour. In Proceedings of the IOP Conference Series: Materials Science and Engineering, International Conference on Chemistry and Material Science (IC2MS) 2017, Malang, Indonesia, 4–5 November 2017; Volume 299. [CrossRef]

15. ASTM International, Standard Test Methods for Chemical Analysis of Hydraulic Cement[1]. 2004; Volume i, 1–31.
16. ASTM International, Standard Test Method for Density , Relative Density (Specific Gravity), and Absorption. 2004; 1–6.
17. ASTM International, Standard Specification for Concrete Aggregates. 2001; Volume 04.
18. ASTM International, Standard Test Method for Flow of Hydraulic Cement Mortar. 2013; 1–2. [CrossRef]
19. ASTM International, Standard Test Method for Compressive Strength of Hydraulic Cement Mortars. 2005; Volume 04, 1–6.
20. ASTM International, Standard Test Method for Splitting Tensile Strength of Cylindrical Concrete Specimens. 2006; 4–8.
21. ASTM International, Standard Test Method for Flexural Strength of Hydraulic-Cement Mortars. 1998; Volume 04, 2–7.
22. Youness, D.; Yahia, A.; Tagnit-Hamou, A. Coupled rheo-physical effects of blended cementitious materials on wet packing and flow properties of inert suspensions. *Constr. Build. Mater.* **2020**. [CrossRef]
23. Estephane, P.; Garboczi, E.J.; Bullard, J.W.; Wallevik, O.H. Three-dimensional shape characterization of fine sands and the influence of particle shape on the packing and workability of mortars. *Cem. Concr. Compos.* **2019**, *97*, 125–142. [CrossRef]
24. Wu, W.; Zhang, W.; Ma, G. Mechanical properties of copper slag reinforced concrete under dynamic compression. *Constr. Build. Mater.* **2010**, *24*, 910–917. [CrossRef]
25. Abdel-Magid, T.I.M.; Hamdan, R.M.; Abdelgader, A.A.B.; Omer, M.E.A.; Ahmed, N.M.R.A. Effect of Magnetized Water on Workability and Compressive Strength of Concrete. *Procedia Eng.* **2017**, *193*, 494–500. [CrossRef]
26. Praveen Kumar, K.; Radhakrishna, X.X. Workability Strength and Elastic Properties of Cement Mortar with Pond Ash as Fine Aggregates. *Mater. Today Proc.* **2020**, *24*, 1626–1633. [CrossRef]
27. Westerholm, M.; Lagerblad, B.; Silfwerbrand, J.; Forssberg, E. Influence of fine aggregate characteristics on the rheological properties of mortars. *Cem. Concr. Compos.* **2008**, *30*, 274–282. [CrossRef]
28. Cu, Y.T.H.; Tran, M.V.; Ho, C.H.; Nguyen, P.H. Relationship between workability and rheological parameters of self-compacting concrete used for vertical pump up to supertall buildings. *J. Build. Eng.* **2020**, *32*. [CrossRef]
29. Pauzi, N.N.M.; Jamil, M.; Hamid, R.; Abdin, A.Z.; Zain, M.F.M. Influence of spherical and crushed waste Cathode-Ray Tube (CRT) glass on lead (Pb) leaching and mechanical properties of concrete. *J. Build. Eng.* **2019**, *21*, 421–428. [CrossRef]
30. Edwin, R.S.; Gruyaert, E.; De Belie, N. Influence of intensive vacuum mixing and heat treatment on compressive strength and microstructure of reactive powder concrete incorporating secondary copper slag as supplementary cementitious material. *Constr. Build. Mater.* **2017**, *155*, 400–412. [CrossRef]
31. ASTM Standard Test Method for Potential Alkali Reactivity of Cement-Aggregate Combinations (Mortar-Bar Method). *Annu. B ASTM Stand.* **2003**, *i*, 1–5.
32. Piasta, W.; Zarzycki, B. The effect of cement paste volume and w/c ratio on shrinkage strain, water absorption and compressive strength of high performance concrete. *Constr. Build. Mater.* **2017**, *140*, 395–402. [CrossRef]
33. Rahmani, K.; Shamsai, A.; Saghafian, B.; Peroti, S. Effect of Water and Cement Ratio on Compressive Strength and Abrasion of Microsilica Concrete. *Middle East J. Sci. Res.* **2012**, *12*, 1056–1061. [CrossRef]
34. Sipil, T.; Teknik, F.; Tengah, S.; Teknik, F.; Tengah, S. The effect of coarse aggregate hardness on the fracture toughness and compressive strength of concrete. *MATEC Web Conf.* **2019**, *258*, 04011.
35. Ouda, A.S.; Abdel-gawwad, H.A. The effect of replacing sand by iron slag on physical, mechanical and radiological properties of cement mortar. *HBRC J.* **2017**, *13*, 255–261. [CrossRef]
36. Guo, Y.; Xie, J.; Li, J. Effect of steel slag as fine aggregate on static and impact behaviors of concrete. *Constr. Build. Mater.* **2018**, *192*, 194–201. [CrossRef]
37. Li, L.G.; Lin, C.J.; Chen, G.M.; Kwan, A.K.H.; Jiang, T. Effects of packing on compressive behaviour of recycled aggregate concrete. *Constr. Build. Mater.* **2017**, *157*, 757–777. [CrossRef]
38. Waheed, A. Properties of Concrete Containing Effective Microorganisms Using Tin Slag as Fine Aggregate Replacement. Master's Thesis, Universiti Teknologi Malaysia, Johor, Malaysia, 2018.
39. Maitra, S.R.; Reddy, K.S.; Ramachandra, L.S. Load Transfer Characteristics of Aggregate Interlocking. *J. Transp. Eng.* **2010**. [CrossRef]
40. Darwin, D.; Kozul, R. *Effects of Aggregate Type, Size, and Content on Concrete Strength and Fracture Energy*; SM Report No. 43, University of Kansas Center for Research, Inc.: Lawrence, KS, USA, 1997.

Article

Physical and Mechanical Properties of Sustainable Hydraulic Mortar Based on Marble Slurry with Waste Glass

Bartolomeo Megna [1,2], Dionisio Badagliacco [1], Carmelo Sanfilippo [1,*] and Antonino Valenza [1]

1   Department of Engineering, University of Palermo, 90128 Palermo, Italy; bartolomeo.megna@unipa.it (B.M.); dionisio.badagliacco@unipa.it (D.B.); antonino.valenza@unipa.it (A.V.)
2   INSTM Research Unity of Palermo, 90128 Palermo, Italy
*   Correspondence: carmelo.sanfilippo01@unipa.it

Abstract: This paper aims to propose and characterize a sustainable hydraulic mortar entirely obtained by the reuse of waste materials, with marble slurry coming from quarries in the northwestern Sicily and glass powder coming from a waste collection plant in Marsala (Province of Trapani). The first was used as raw material to produce the mortar binder by a kilning and slaking process, while the second was used as a pozzolanic additive. The chemical and morphological characterization of the marble slurry was done by XRD, FTIR, STA and SEM analyses. Glass powder was analyzed through particle size distribution measurements, XRD and standard pozzolanic tests. Hydraulic mortars constituted by slaked lime from kilned marble slurry and waste glass powder (LGS) were prepared beside commercial Natural Hydraulic Lime (NHL) based mortars (NGS) and air-hardening lime (LSS)-based mortars. Mechanical and absorption properties of the mortars were investigated as a function of the grain size of the glass powder by means of three-point bending and compressive strength tests, capillary uptake, helium pycnometry and simultaneous thermal analysis. The results demonstrated that the formulation LGS exhibits significantly improved mechanical and absorption properties compared to air-hardening mortars (LSS). It confirms the possibility of producing a more sustainable hydraulic mortar exclusively from waste materials for civil engineering.

Keywords: hydraulic mortars; waste materials; pozzolanic aggregates; recycled glass; marble slurry

Citation: Megna, B.; Badagliacco, D.; Sanfilippo, C.; Valenza, A. Physical and Mechanical Properties of Sustainable Hydraulic Mortar Based on Marble Slurry with Waste Glass. *Recycling* **2021**, *6*, 37. https://doi.org/10.3390/recycling6020037

Academic Editors: José Neves and Ana Cristina Freire

Received: 8 April 2021
Accepted: 4 June 2021
Published: 9 June 2021

1. Introduction

In the last decades, a considerable amount of research has concerned the use of waste marble [1–7] and waste glass powders [8–18], separately, as additives to improve the physical and mechanical properties of mortars and concretes. However, the possibility to use these two materials simultaneously to produce a new sustainable binding system has not been investigated to date.

Waste materials causes a tremendous effect on air, water, vegetation, animals, human health and living conditions [19]. Overall, it is predicted that the 12 billion tons of waste produced in 2002 (with 11 billion tons corresponding to industrial waste) will increase up to 25 billion tons in 2025 [20,21]. It is evident that the increase will pose a tremendous threat to environmental health in the absence of appropriate measures.

It is well known that Portland cement production is one of the most environmental unfriendly process due to its high working temperature (1400–1500 °C) and the massive consumption of raw materials such as limestone and clay. Cement production causes a considerable amount of greenhouse gas emissions both by carbonate decomposition and by fossil fuels combustion to heat the kiln. The carbon dioxide to Portland cement weight ratio is close to 1:1 [22]. Moreover, the extraction of raw materials used for its production is further an environmental issue and source of CO_2 emissions [23].

Marble slurry has to be disposed as hazardous industrial waste [24], according to the limits established by Italian law. If not properly disposed, it is generally poured into riverbeds, damaging aquatic populations. Since slurry mainly consists of $CaCO_3$, it could be used as raw material for lime production. This would result in a lowering of the environmental impact of lime production due to the reduction of energy consumption for the extraction and grinding of the raw materials, which contributes to more than 10% of the CO_2 emissions associated with binder production [25,26].

The issue of management and disposal of sludge from the exploitation of carbonatic rocks is of increasing interest. In particular, in the area of Custonaci (Province of Trapani), the marble quarries produce approximately 200,000 tons/year of sawing mud, which is mostly disposed of by filling the same quarries, demonstrating the vast abundance of waste, and encouraging its possible use on an industrial scale [24].

The literature is also focused on the possible use of waste marble for production of clay bricks [8,27–29].

Glass is a recyclable material that can be indefinitely recycled [9]. Even though a portion of this glass is easily recycled in the glass manufacturing industry, not all used glass can be recycled into new glass because of impurities, cost or mixed colors. Therefore, several decades ago, research was started to investigate the possibility of using waste glass in concrete production [18]. As noted by [10,11], waste glass can be used for many applications, such as the production of binders such as a siliceous aggregate for cement and concretes. Refs. [12,13] used glass powder from flat glass waste as pozzolanic material in the preparation of hydraulic limes or hydraulic mortars. According to [14], the amorphous silica (SiO_2) reacts with the portlandite generated during cement hydration and forms gels of calcium silicate hydrate (C–S–H). The effect of glass grain size incorporated into cement-based products was investigated by [30], who verified that fine glass powders improve the properties of concrete, while coarse aggregates are generally harmful. Ref. [31] and described the preparation of lime mortars in which the sand was replaced by crushed glass and calcium carbonate aggregates. The pozzolanic properties of the glass are first notable at particle sizes below 300 μm [32]. It was argued by [33] that under 100 μm, the glass might exhibit higher pozzolanic reactivity compared to fly ash at low percentages of cement replacement after 90 days of curing. Previous studies observed a strong reaction between alkali materials in the cement and reactive silica, commonly referred to as the alkali-silica reaction (ASR). The grain size of the glass powder may influence ASR expansion. However, different authors claimed controversial results with ASR expansion increases [34] or decreases [35] with the fineness of the particles.

This paper highlights the possibility of using waste marble slurry as raw material to produce calcitic lime, and waste glass powder as a pozzolanic aggregate to produce a hydraulic binder entirely based on waste materials. Since the size of the particles strongly influences waste glass reactivity, three different grain size distributions were investigated. Mechanical and absorption properties of the new ecocompatible hydraulic binding system were compared with air-hardening lime and natural hydraulic lime. The experimental results provide evidence that the proposed hydraulic mortars can be a sustainable alternative to cement or natural hydraulic lime-based mortars, especially for nonstructural purposes.

2. Materials and Methods

2.1. Characterization of Raw Materials and Products

The raw materials used in this work came from two different sources of waste. Limestone slurry came from Custonaci (Province of Trapani), northwestern Sicily, basin quarries and resulted from the sawing-cooling system and the polishing system of limestone blocks. The amount of slurry produced in this site is around 2×10^5 tons/year [24]. Quicklime was obtained by kilning the marble slurry in a muffle furnace at 900 °C for 24 h. Slaked lime was obtained by mixing quicklime and distilled water in a ratio of 1:2.5 by weight. After ten days an almost complete conversion of Calcium oxide to calcium hydroxide occurred and, hence, mortars were manufactured. The waste glass powder was provided by a selective waste collection site located in Marsala (Province of Trapani) operated by Sarco S.r.l. Finally, common river sand, consisting mostly of quartz with traces of biotite and feldspar, was used as aggregate of the mortars.

The characterization of raw materials and products was made by means of X-Ray Diffractometry (XRD), Simultaneous Thermal Analysis (STA), Low Vacuum Scanning electron microscopy (SEM) with Energy Dispersion X-Ray Spectroscopy (EDS) and Fourier Transformed Infrared Spectroscopy (FTIR). Grain size distribution of glass powder was evaluated using standard sieves with a kilogram of glass powder.

According to the supplier, the granulometric range of the used sand was from 0.125 to 5.6 mm.

XRD analysis was performed with a Panalytical Empyrean Diffractometer equipped with a PixCel 1D detector with Cu Ka radiation in the 2θ range 6–60°.

Simultaneous Thermal Analysis was conducted in a Netzsch Jupiter F1 STA 449, in the range 30–1000 °C to evaluate the pozzolanic behavior of the glass powder [36].

The ESEM used in this work was a FEI Quanta 200 FEG coupled with EDS elemental analysis. The observation was performed using a low vacuum mode in order to analyze the samples without any further preparation.

FTIR spectroscopy was performed on a Shimadzu FTIR 8000, in the wavenumber range 400–4000 cm^{-1} with a 4 cm^{-1} spectral resolution. In order to obtain information on the organic compounds, the latter was extracted by submerging marble slurry in different suitable solvents, i.e., acetone and ethylic alcohol, and after an acid attack with a 5% by weight solution of hydrochloric acid. In both cases, the solid phase was mixed with reagent grade potassium bromide in a 1/100 weight ratio.

Pozzolanicity tests (according to UNI 11471, January 2013 [37]) were performed to establish the pozzolanic activity of each selected range of grainsize glass powder:

- $D < 125$ (μm).
- $125 < D < 210$ (μm).
- $350 < D < 500$ (μm).

Finally, thin sections of mortars were observed by optical microscopy to identify the presence of reaction borders.

2.1.1. Waste Glass Powder

The grain size distribution of the glass powder coming from the recycling plant is shown in Table 1. A significant percentage of the glass powder had diameters greater than 500 μm.

Table 1. Grain size distribution of glass powder.

Sieve Passing Diameter [μm]	Percentage Weight Retained [%]
500	36.2
350	11.4
210	19.3
125	10.3
53	20.2
0	2.3
Loss on Measurement	0.3

This result suggests the benefit of grinding the glass powder to increase the amount of smaller particles fraction characterized by higher reactivity [36,38].

The X-Ray test shown in Figure 1 (above) confirms the amorphous nature of the glass powder, mainly consisting of silicon, calcium and sodium oxides, as also highlighted by EDS analysis shown in Figure 1 (below). The same composition was also obtained by [39] who highlighted the possibility of glass powder acting as a pozzolanic or even a cementitious material.

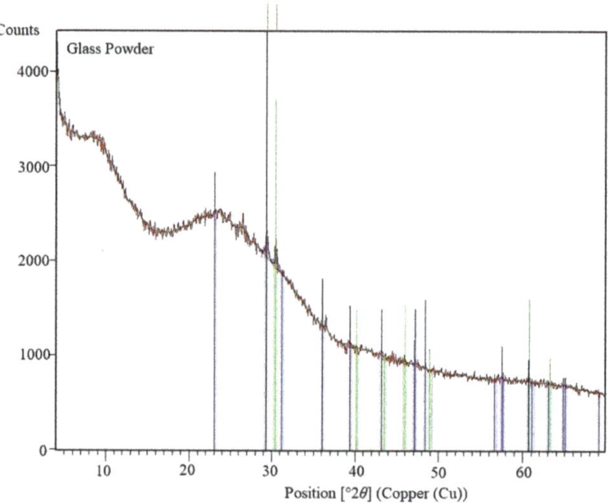

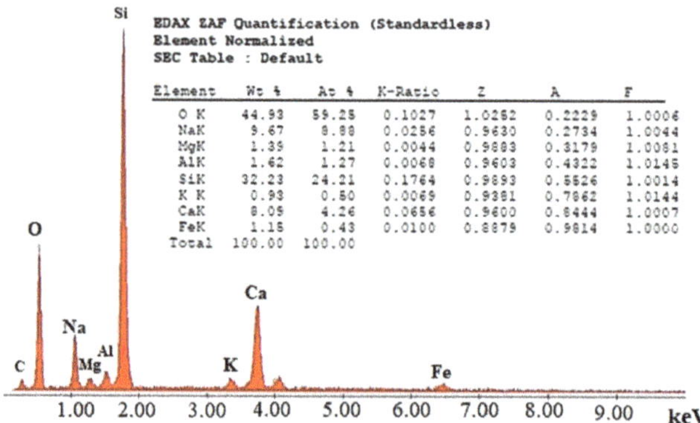

Figure 1. X-Ray pattern of glass powder (**above**); elemental Analysis of the glass powder (**below**).

The SEM micrographs in Figure 2 show as-received waste glass powders and ground waste glass powders, highlighting their angular shape.

The results of pozzolanic activity tests are shown in Figure 3.

From Figure 3, it is evident that each glass grain size showed pozzolanic behavior after 8 days. After 15 days, finer grain sizes exhibited higher reactivity, while coarser powders had lower reactivity as indicated by the small distance of the representing point from the equilibrium curve. This result is in accordance to that obtained by [33] who recorded a higher pozzolanic reactivity of fine glass powders below 100 μm compared to fly ash. According to these considerations, mortars with coarser glass grain sizes were prepared using a grain size distribution from 210 to 350 μm instead of 350 to 500 μm as in pozzolanic tests.

2.1.2. Marble Slurry

The XRD analysis shown in Figure 4 (above) confirmed that marble slurry consists of $CaCO_3$ (Calcite), quantified as 98% according to the STA results Figure 4 (below). It is noticeable that a mass loss occurred between 330–360 °C due to organic compound decomposition, i.e., the resin used for the puttying of marble slates as also identified by FTIR analysis.

Figure 2. SEM micrographs of as-received (**above**) and ground (**below**) glass powder.

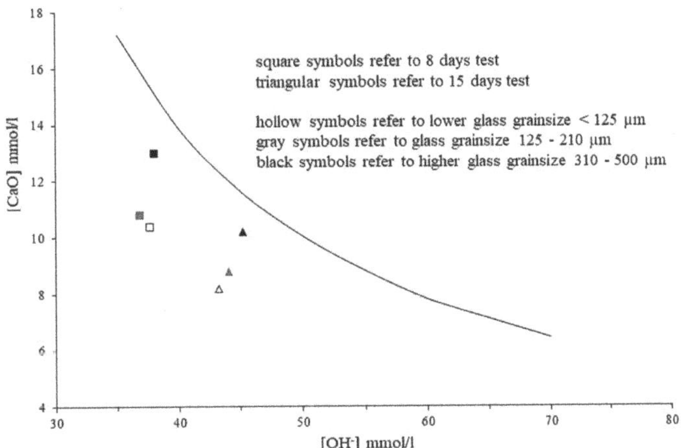

Figure 3. Pozzolanic test results.

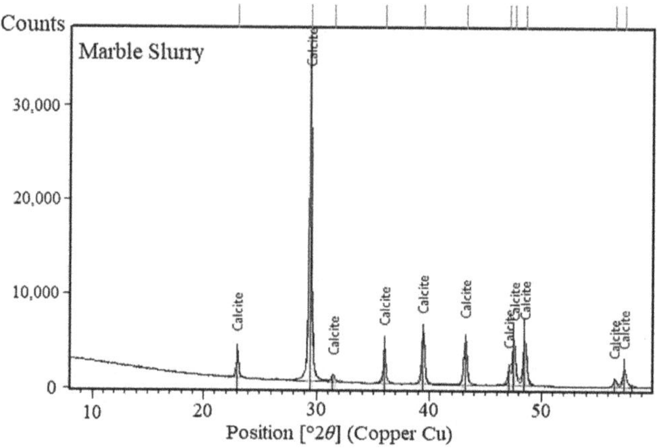

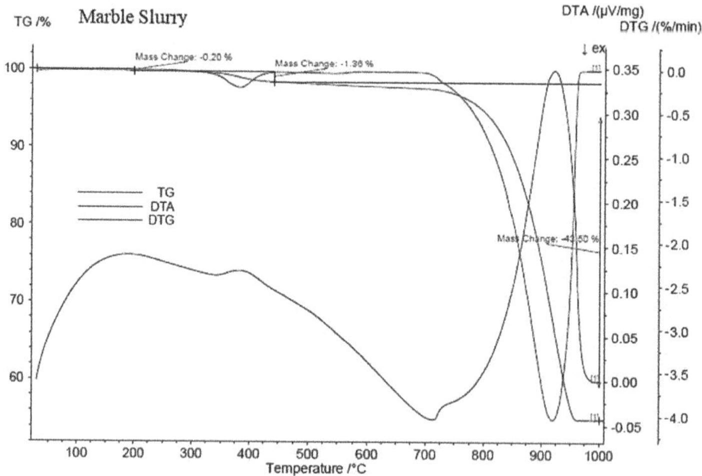

Figure 4. Characterization of marble slurry: XRD pattern (**above**); STA results (**below**).

FT-IR analysis, shown in Figure 5, was performed after solvent extraction using ethanol or acetone, and on the residue of dissolution with hydrochloric acid. The latter procedure provided the best results and confirmed that the organic compound was a cured polystyrene resin.

2.1.3. Kilned Marble Slurry and Slaked Lime

X-ray diffraction analyses were conducted on the kilned marble slurry and slaked lime (Figure 6).

As expected, XRD patterns highlight the conversion of marble slurry to lime and portlandite after kiln heating and slaking, respectively. STA performed on slaked lime (Figure 7) indicates that after 10 days an almost complete conversion of Calcium Oxide to Calcium Hydroxide occurred.

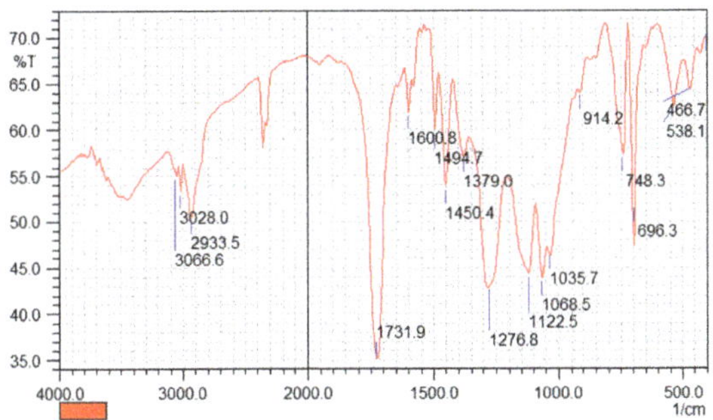

Figure 5. Spectrum of HCl attack residue from marble (the solvent was HCl).

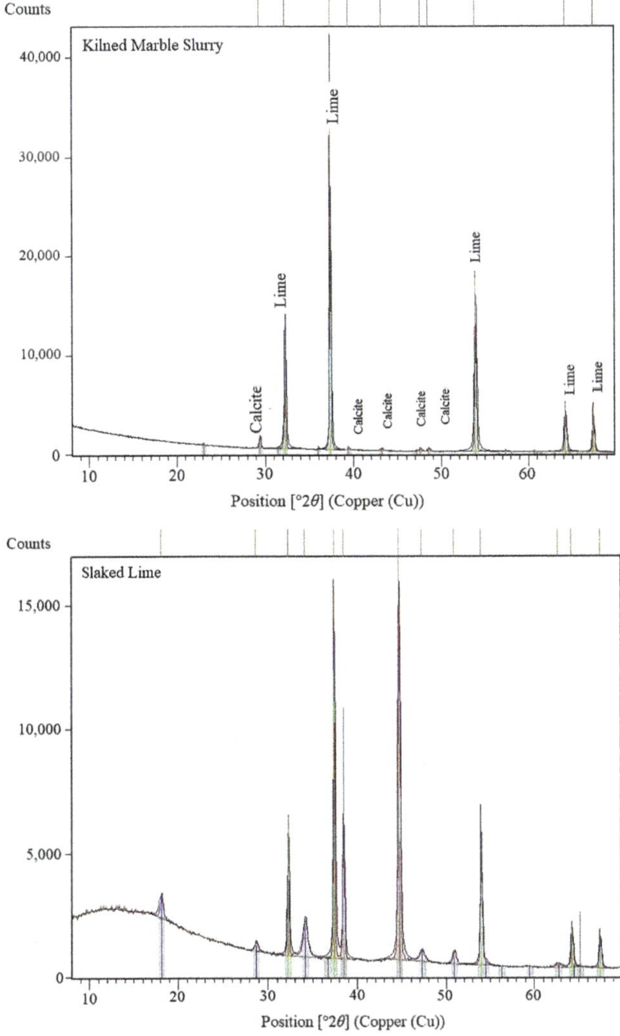

Figure 6. X-Ray diffractograms of kilned marble slurry (**above**) and slaked lime (**below**).

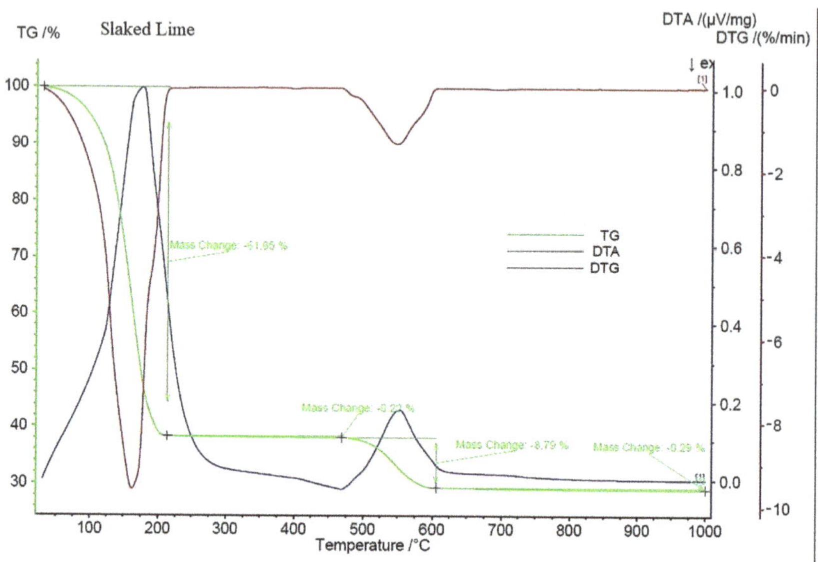

Figure 7. Simultaneous Thermal Analysis (STA) on slaked lime.

The use of a Scanning Electron Microscope (SEM-ESEM) combined with EDS microscopy was performed for kilned marble slurry mixed with water. Figure 8 highlights the predominance of hexagonal crystals and the dimensions of each component:

- Order of tens of microns for the aggregates.
- Under one micron for the crystals.

2.2. Types of Proposed Mortars

In this work, nine different types of mortars were prepared. Three were constituted by slaked lime from marble slurry (L), glass powder (G) and river sand (S) in 1:1:1.5 volume ratio with three different glass grain size, i.e., lower than 125 microns (a), between 125 and 210 microns (b) and between 210 and 350 microns (c). Three were composed of Natural Hydraulic Lime (N) with a water to binder weight ratio equal to 0.8, glass powder (G) and river sand (S) in 1:1:1.5 volume ratio with the three different glass grain sizes as above. The last three were obtained by mixing slaked lime from marble slurry (L) and river sand (S) used in two different fractions; the first characterized by the same grain size distribution of the glass powder in a 1:1 volume ratio with the slaked lime and the second used as received in 1:1.5 volume ratio with the slaked lime. This composition was chosen to reproduce the aggregate grain size distribution of hydraulic mortars in air-hardening mortars. The mortars were named as shown in Table 2, according to the composition.

LGS formulations were aimed at evaluating the reactivity of recycled glass powder with calcium hydroxide ($Ca(OH)_2$) as a function of the grain size. The LSS and NGS mixes were prepared with the same composition of the LGS mixes to evaluate the performance of the hydraulic binder compared to air-hardening lime and natural hydraulic lime.

The samples were prepared by mixing the dry powders in a container for at least 2 min to homogenize the constituents, then the slaked lime (for LGS and LSS mortars) or water (for NGS mortars) were added, and the components were further mixed for at least 4 min until a homogeneous and workable mixture was obtained.

The mortars were then put in 40 × 40 × 160 mm molds for the three-point bending and compressive tests according to the standard UNI EN 1015-11 (2013) [40]. The specimens were cured in the molds for five days in laboratory conditions (22 ± 2 °C and 50 ± 5% RH) in order to allow the samples to reach adequate mechanical properties to facilitate the demolding process without causing damages. After five days, the samples were removed from the molds and placed in a climatic chamber at 22 ± 2 °C and 90 ± 5% RH for three months. A high relative humidity was chosen to favor hydraulic hardening with respect to air-hardening for the proposed hydraulic binder-based mortars (LGS), while a long curing time (3 months) was chosen to allow the air-lime (LSS) to reach adequate hardening.

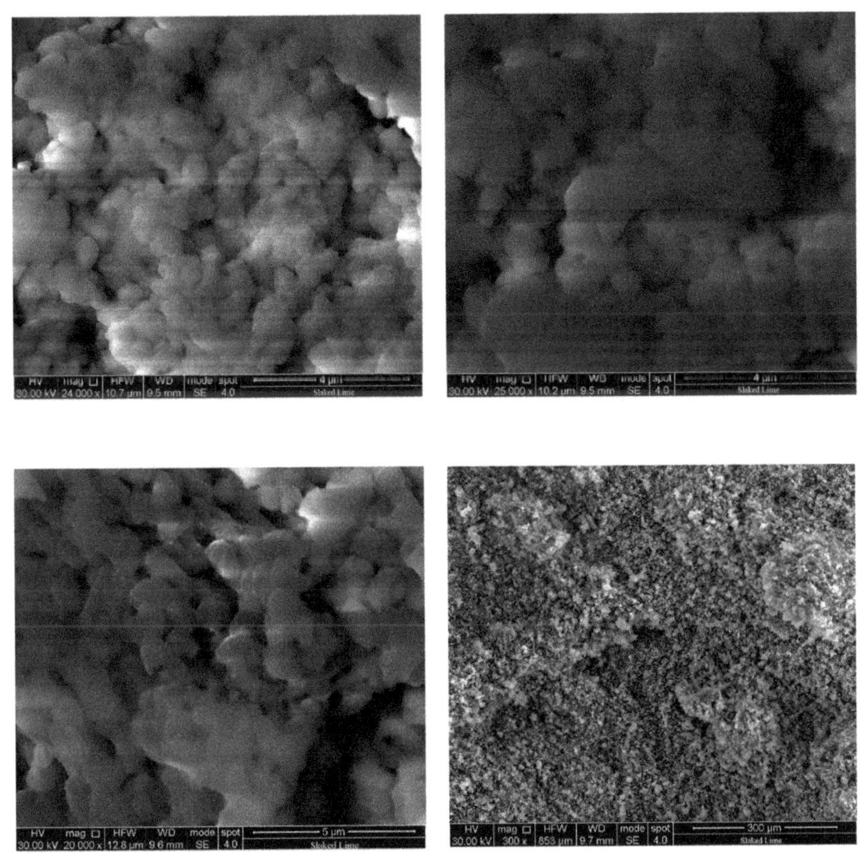

Figure 8. SEM micrographs of slaked lime.

Table 2. Composition by volume of the mortars: a = lower than 125 μm, b = 125 ÷ 210 μm; c = 210 ÷ 350 μm.

Type of Mortars	Binder		Glass Powder Grain Size			River Sand Grain Size			River Sand
	Slaked Lime	NHL 3,5	a	b	c	a	b	c	As Received
LGS_1	1		1						1.5
LGS_2	1			1					1.5
LGS_3	1				1				1.5
NGS_1		1	1						1.5
NGS_2		1		1					1.5
NGS_3		1			1				1.5
LSS_1	1					1			1.5
LSS_2	1						1		1.5
LSS_3	1							1	1.5

2.3. Experimental Characterization of the Mortars

2.3.1. Three-Point Bending Test

For each type of mortar, five specimens were tested at 90 days of curing, and the results were recorded as the average and standard deviation of the tensions at breakage obtained from valid tests (at least three specimens for each type). Flexural strength was obtained for each sample by the following equation:

$$\sigma_{max} = \frac{3PL}{2bh^2} \tag{1}$$

The tests were carried out according to the standard, UNI EN 1015-11 (2013) [40] in force control in a Zwick/Roell Z005 universal electromechanical machine equipped with a 5 kN load cell. The span value was set to 100 mm, the preload to 20 N and the loading speed to 20 N/s in order to obtain the breakage of the specimens between 15 and 60 s.

2.3.2. Compressive Test

For each type of mortar, five specimens were tested at 90 days of curing, and the results were recorded as average and standard deviation of the maximum compressive strength obtained from valid tests (at least three specimens for each type). The tests were carried out in force control on halves of the samples previously tested for bending with reference to the standard UNI EN1015-11 [40], in a universal electromechanical machine MP Strumenti Tools WANCE UTM 502, at a loading speed equal to 50 N/s to obtain the rupture for all the specimens between 20 and 90 s.

2.3.3. Porosity Measurement

For each type of mortar, the apparent densities (from the ratio between the weight and the apparent volume of each specimen) and the real densities (by performing at least two tests of helium picnometry on fragments of samples) were calculated at 90 days of curing. The porosity (or empty index) was evaluated using the following equation:

$$Porosity \: [\%] = 100 \left(1 - \frac{\rho_{app}}{\rho_{real}} \right) \tag{2}$$

2.3.4. Water Absorption and Desorption Test

Absorption and desorption tests were conducted at 90 days of curing on at least three specimens for each type of mortar investigated on halves of the samples previously tested for bending, according to the standard UNI EN 1015-18 (2004) [41]. Before performing the water absorption test, the samples were dried in an oven at 60 °C until constant mass was reached. The samples were then placed in a container with the fracture surface of the prism downward on suitable supports to maximize the contact between the water and the surfaces of the samples. The water level in the container was maintained between 5 and 10 mm throughout the test, which was performed in laboratory conditions 22 ± 2 °C and 50 ± 5% RH. The mass of each sample was recorded after 0, 10, 90 min of immersion, and the absorption coefficient was evaluated by the following equation:

$$C = 0.1(M_2 - M_1) \tag{3}$$

where: C is the absorption coefficient, M_1 is the mass of the sample after 10 min of immersion and M_2 is the mass of the sample after 90 min of immersion. On completion of the test, the specimens were left in the water container until saturation. Water desorption curves were obtained by monitoring at regular intervals (24 h) the weight loss of the mortars at 22 ± 2 °C and 50 ± 5% RH.

2.3.5. Simultaneous Thermal Analysis (STA)

Simultaneous thermal analysis was performed using a Netzsch STA 449 Jupiter F1 instrument. The tests were carried out in the 30–1000 °C range with a heating rate of 10 °C/min and nitrogen flux of 20 mL/min. Thermogravimetry is considered one of the most effective analytical techniques for determining mortar hydraulicity [42,43]. Recorded weight losses were labeled according to the temperature range: 30–200 °C as H_2O_{abs}; 200–600 °C as H_2O_{idr}; above 600 °C as CO_2.

3. Results and Discussion

3.1. Thin Sections Observation

Mortars samples were prepared to produce thin sections for petrographic characterization after one month of hardening. Particularly, thin section observation is a powerful tool to identify the presence of reaction borders between lime and pozzolanic aggregate [44].

Figure 9 shows the presence of small reaction fringes, visible all around some of the glass grains and indicated with red arrows, after only one month of hardening, which was considered a good sign of glass pozzolanic behavior.

Figure 9. Microscopic images in observations with a crossed-polarized lens, (**A,C**), and parallel polarized lens, (**B,D**).

3.2. Three-Point Bending Test

The results obtained from the bending tests are shown in Figure 10a. The values of flexural strength of the proposed material (LGS) were clearly higher than those obtained by simple air-hardening lime mortars (LSS), and particularly for the lowest grain size particles, the LGS mortars reached more than half of the NGS mortars. In fact, the effect of the grain size was remarkable, leading to a significant increase in the flexural strength of approximately 100% between LGS_3 and LGS_1. These results are in agreement with those obtained by [36,45,46] who evidenced an increase in the flexural behavior when recycled glass powder was used in concrete, both in substitution of cement and as a fine aggregate.

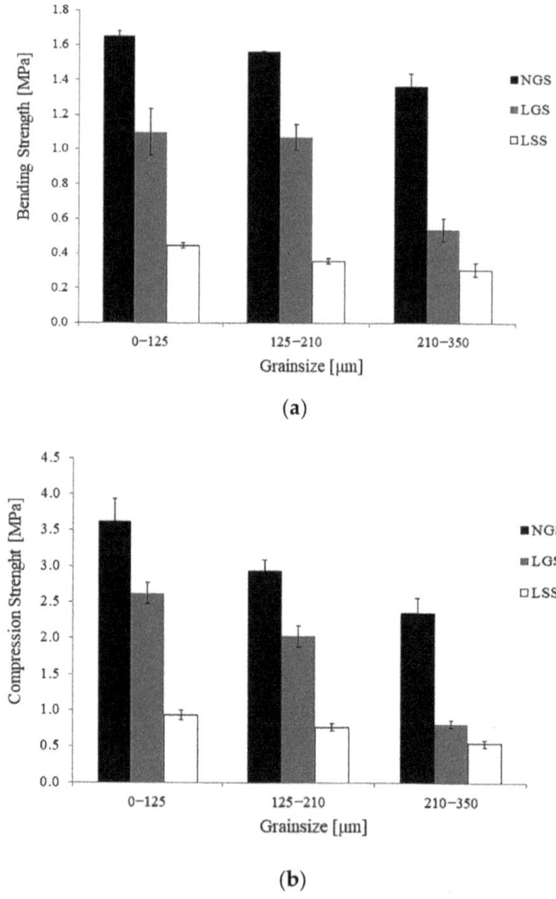

Figure 10. (**a**) Three Point Bending, and (**b**) compression results.

3.3. Compressive Test

As obtained for bending, the compressive test results (Figure 10b) showed resistance values of the proposed material (LGS) significantly higher than those exhibited by air lime-based mortars (LGS) and similar to those obtained by natural hydraulic lime-based mortars (NGS). Both compressive and bending results confirmed the higher reactivity of glass with lower particle size; hence, it is preferable to mill the raw material provided by the recycling plant to improve its performance as a pozzolanic additive. It is also important to note that several studies reported that, although early-age strength can be lower compared to the reference when recycled waste glass is used in concrete, later-age strength may increase [47,48]. For this reason, it is plausible to expect that the compressive strength of the LGS mortars at later curing times may further increase not only for the progress of the carbonation process but also for the continuation of the pozzolanic reactions between the glass powders and the residual portlandite.

3.4. Porosity Test

From the data reported in Table 3, it can be assessed that LGS mortars are characterized by higher porosity compared to natural hydraulic lime-based mortars (NGS). This can be a sign of prospective better thermal insulation performance, which can further reduce the operating costs of these plasters.

Table 3. Porosity of samples calculated by density measurements.

	Grain Size [μm]	Porosity [%]
	0–125	32
NGS	125–210	32
	210–350	33
	0–125	37
LGS	125–210	37
	210–350	38
	0–125	35
LSS	125–210	34
	210–350	35

3.5. Water Absorption and Desorption Test

The results for absorption coefficients are reported in Figure 11. The lowest values were once again reached for LGS_a and LGS_b mortars. In this case, performances were even higher than for NGS samples, which confirms the possibility of using the LGS mortars in place of NGS for civil engineering.

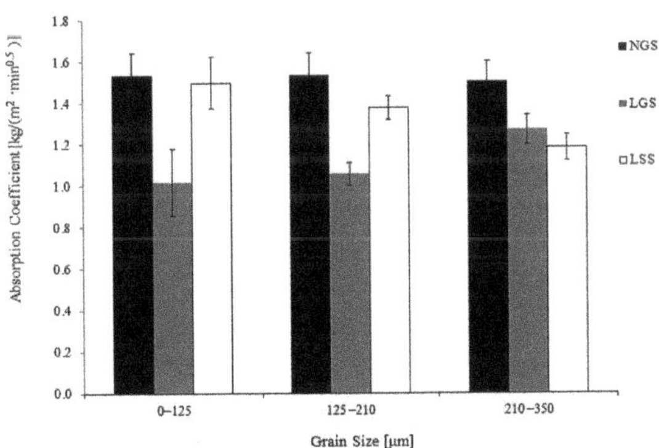

Figure 11. Water absorption coefficient results.

Data obtained for porosity and water absorption coefficients were confirmed by the water absorption and desorption curves (Figures 12 and 13), which revealed slower absorption and faster evaporation of water for the LGS samples compared to the NGS samples. The lower water absorption coefficient of the LGS mortars could be due to the higher average dimensions of the pores compared to the air lime mortars, leading to a slower water absorption rate due to capillary action. This hypothesis was validated by observing water absorption curves showing that after 24 h the mortars LGS125 and LGS210 did not reach saturation, unlike the other mortars. This property could be considered the main advantage for manufacturing plasters.

3.6. Simultaneous Thermal Analysis

By comparing the results of the Simultaneous Thermal Analysis in Figure 14 it is possible to assess the hydraulicity of LGS mortars. In particular, that the weight loss associated with the presence of calcium hydroxide relative to the LSS mortars was considerably higher than that which occurred for LGS mortars. This is because the glass powder used as aggregate reacts with lime to produce a calcium silicate hydrate. Furthermore, the effect of grain size on the reactive efficiency of the glass powder was evident since lower values of the calcium hydroxide peak were recorded at lower granulometry due to the higher surface/volume ratio of the particles.

According to [43], the classification of mortar hydraulicity can be performed more effectively by dividing the value of CO_2 and H_2Oidr associated weight loss by absorbed water, H_2Oabs. The resulting graph in Figure 15, clearly shows the different hydraulic behavior of LGS mortars compared to LSS mortars, as the representing points of the LGS are closer to NHL mortars ones.

The mechanical performances were in good agreement with the thermogravimetric characterization, as clearly shown in Figure 16.

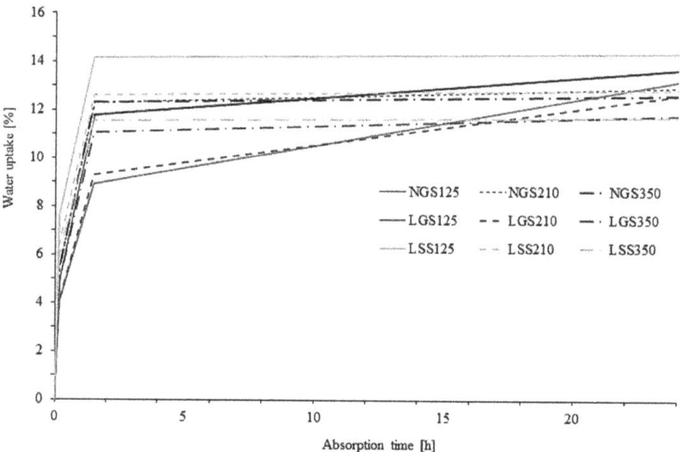

Figure 12. Water absorption curves.

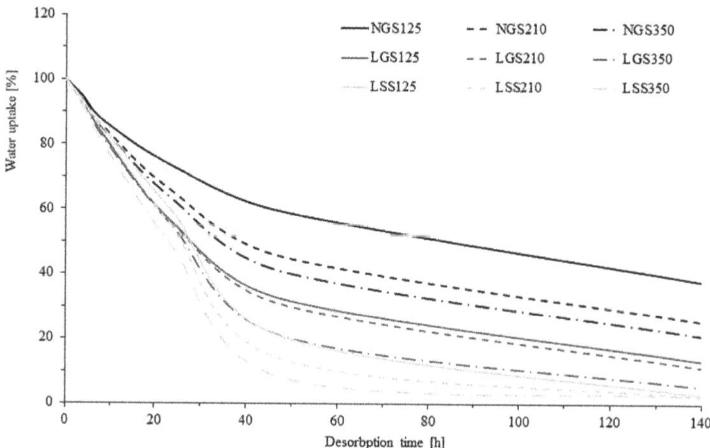

Figure 13. Water desorption curves.

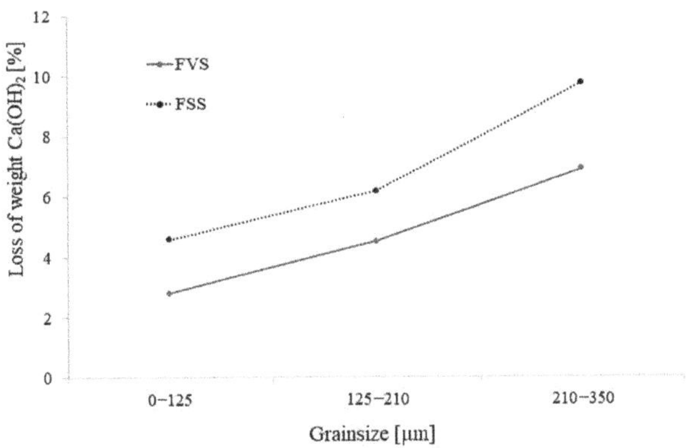

Figure 14. Weight loss associated with portlandite obtained from STA.

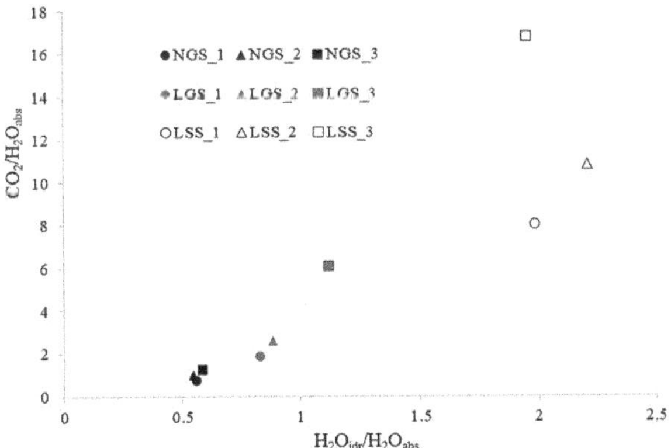

Figure 15. Correlation between carbon dioxide and hydraulic water obtained from STA.

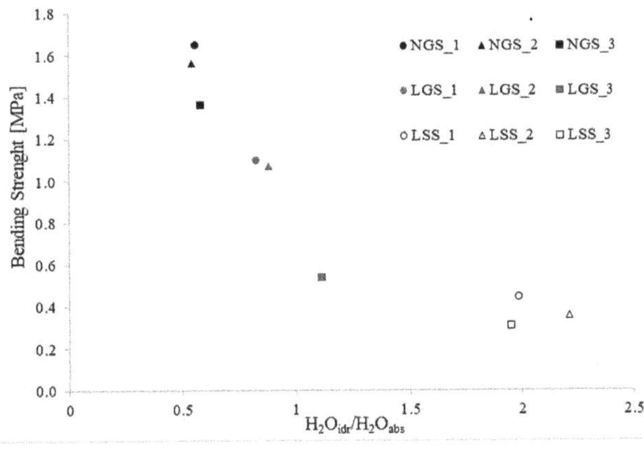

Figure 16. Correlation between bending strength and hydraulic water.

4. Conclusions

This work proposes a more sustainable binder produced using wastes as raw materials, reducing the environmental issues of both binder production and waste disposal.

The preliminary characterization of the materials revealed the opportunity to use marble slurry to produce an air-hardening lime that can act as a hydraulic binder when mixed with glass powder obtained from urban waste collection. Thermal analysis and microscopic observation confirmed the hydraulic behavior of the binder. Moreover, mechanical tests clearly showed good mechanical performance of the hydraulic binder based on air-hardening lime with better performance at lower grain size of the glass powder. Particularly, mechanical properties of mortars with fine glass powder were considerably higher than air-hardening mortars, while the differences were reduced using coarser glass particles.

It is fundamental to note that pozzolanic behavior of the glass powder could be obtained if the fineness did not exceed 210 μm. The experimental data showed that this system, which exploits the coupling of two waste materials, their transformation and combined use, can be used to obtain mortars with high mechanical performance. Finally, the proposed mortars showed higher porosity compared to natural hydraulic lime mortars, and comparable mechanical properties. These results indicate that these kinds of mortars may offer a further advantage related to their better insulating properties.

5. Patents

The content is object of Italian Patent Application No. 102018000007810 filed on 3 August 2018.

Author Contributions: Conceptualization, B.M. and A.V.; methodology, B.M. and D.B.; validation, B.M., D.B. and A.V.; formal analysis, B.M. and D.B.; investigation, B.M., D.B. and C.S.; resources, B.M. and A.V.; data curation, B.M., D.B. and C.S.; writing—original draft preparation, D.B.; writing—review and editing, D.B. and B.M.; supervision, A.V.; project administration, A.V. All authors have read and agreed to the published version of the manuscript.

Funding: This research received no external funding.

Data Availability Statement: All data is contained within the article.

Conflicts of Interest: The authors declare no conflict of interest.

References

1. Kabeer, K.I.S.A.; Vyas, A.K. Experimental investigation on utilization of dried marble slurry as fine aggregate in lean masonry mortars. *J. Build. Eng.* **2019**, *23*, 185–192. [CrossRef]
2. Ashish, D.K. Feasibility of waste marble powder in concrete as partial substitution of cement and sand amalgam for sustainable growth. *J. Build. Eng.* **2018**, *15*, 236–242. [CrossRef]
3. Rana, A.; Kalla, P.; Csetenyi, L.J. Sustainable use of marble slurry in concrete. *J. Clean. Prod.* **2015**, *94*, 304–311. [CrossRef]
4. Ulubeylia, G.C.; Artir, R. Properties of Hardened Concrete Produced by Waste Marble Powder. *Procedia Soc. Behav. Sci.* **2015**, *195*, 2181–2190. [CrossRef]
5. Aliabdo, A.A.; Elmoaty, M.A.; Auda, E.M. Re-use of waste marble dust in the production of cement and concrete. *Constr. Build. Mater.* **2014**, *50*, 28–41. [CrossRef]
6. Corinaldesi, V.; Moriconi, G.; Naik, T.R. Characterization of marble powder for its use in mortar and concrete. *Constr. Build. Mater.* **2010**, *24*, 113–117. [CrossRef]
7. Aruntas, H.Y.; Guru, M.; Dayi, M.; Tekin, I. Utilization of waste marble dust as an additive in cement production. *Mater. Des.* **2010**, *31*, 4039–4042. [CrossRef]
8. Sutcu, M.; Alptekin, H.; Erdogmus, E.; Er, Y.; Gencel, O. Characteristics of fired clay bricks with waste marble powder addition as building materials. *Constr. Build. Mater.* **2015**, *82*, 1–8. [CrossRef]
9. Sobolev, K.; Türker, P.; Soboleva, S.; Iscioglu, G. Utilization of waste glass in ECO cement, strength properties and microstructural observations. *J. Waste Manag.* **2006**, *27*, 971–976. [CrossRef] [PubMed]
10. Shi, C.; Zheng, K. A review on the use of waste glasses in the production of cement and concrete. *Resour. Conserv. Recycl.* **2007**, *52*, 234–247. [CrossRef]
11. Shekhawat, B.S.; Aggarwal, D.V. Utilisation of Waste Glass Powder in Concrete—A Literature Review. *Int. J. Innov. Res. Sci. Eng. Technol.* **2014**, *3*, ISSN: 2319-8753.
12. Fragata, A.; Paiva, H.; Velosa, A.L.; Veiga, M.R.; Ferreira, V.M. *Application of Crushed Glass Residues in Mortars, Sustainable Construction, Materials and Practices: Challenge of the Industry for the New Millennium*; Portugal, S.B., Ed.; IOS Press: Amsterdam, The Netherlands, 2007.

13. Edwards, D.D.; Allen, G.C.; Ball, R.J.; El-Turki, A. Pozzolanic properties of glass fines in lime mortars. *Adv. Appl. Ceram.* **2007**, *106*, 309–313. [CrossRef]
14. Omran, A.; Tagnit-Hamou, A. Performance of glass-powder concrete in field applications. *Constr. Build. Mater.* **2016**, *109*, 84–95. [CrossRef]
15. Ramdani, S.; Guettala, A.; Benmalek, M.L.; Aguiar, J.B. Physical and mechanical performance of concrete made with waste rubber aggregate, glass powder and silica sand powder. *J. Build. Eng.* **2019**, *21*, 302–311. [CrossRef]
16. Olivera, R.; de Brito, J.; Veiga, R. Reduction of the cement content in rendering mortars with fine glass aggregates. *J. Clean. Prod.* **2015**, *95*, 75–88. [CrossRef]
17. Penacho, P.; de Brito, J.; Veiga, M.R. Physico-mechanical and performance characterization of mortars incorporating fine glass waste aggregates. *Cem. Concr. Compos.* **2014**, *50*, 47–59. [CrossRef]
18. Paul, S.C.; Savija, B.; Babafemi, A.J. A comprehensive review on mechanical and durability properties of cement-based materials containing waste recycled glass. *J. Clean. Prod.* **2018**, *198*, 891–906. [CrossRef]
19. Togrul Tunc, E. Recycling of marble waste: A review based on strength of concrete containing marble waste. *J. Environ. Manag.* **2019**, *231*, 86–97. [CrossRef] [PubMed]
20. Yoshizawa, S.; Tanaka, M.; Shekdar, A.V. Global trends in waste generation. In *Recycling, Waste Treatment and Clean Technology*; Gaballah, I., Mishar, B., Solozabal, R., Tanaka, M., Eds.; TMS Mineral, Metals and Materials Publishers: Madrid, Spain, 2004; pp. 1541–1552.
21. Pappu, A.; Saxena, M.; Asolekar, S.R. Solid wastes generation in India and their recycling potential in building materials. *Build. Environ.* **2007**, *42*, 2311–2320. [CrossRef]
22. Hanle, L.; Jayaraman, K.; Smith, J. CO_2 Emissions Profile of the US Cement Industry. Available online: https://www3.epa.gov/ttnchie1/conference/ei13/ghg/hanle.pdf (accessed on 6 June 2021).
23. Ali, M.B.; Saidur, R.; Hossain, M.S. A review on emission analysis in cement industries. *Renew. Sustain. Energy Rev.* **2011**, *15*, 2252–2261. [CrossRef]
24. Rizzo, G.; D'Agostino, F.; Ercoli, L. Problems of soil and groundwater pollution in the disposal of "marble" slurries in NW Sicily. *Environ. Geol.* **2008**, *55*, 929–935. [CrossRef]
25. Halbert, G.; Billard, C.; Rossi, P.; Chen, C.; Roussel, N. Cement production technology improvement compared to factor 4 objectives. *Cem. Concr. Res.* **2010**, *40*, 820–826. [CrossRef]
26. Gartner, E. Industrially interesting approaches to "low-CO_2" cements. *Cem. Concr. Res.* **2004**, *34*, 1489–1498. [CrossRef]
27. Al-Fakih, A.; Mohammed, B.S.; Liew, M.S.; Nikbakht, E. Incorporation of waste materials in the manufacture of masonry bricks: An update review. *J. Build. Eng.* **2019**, *21*, 37–54. [CrossRef]
28. Munir, M.J.; Abbas, S.; Nehdi, M.L.; Kazmi, S.M.S.; Khitab, A. Development of Eco-Friendly Fired Clay Bricks Incorporating Recycled Marble Powder. *J. Mater. Civ. Eng.* **2018**, *30*, 04018069. [CrossRef]
29. Bilgin, N.; Yeprem, H.A.; Arslan, S.; Bilgin, A.; Günay, E.; Marşoglu, M. Use of waste marble powder in brick industry. *Constr. Build. Mater.* **2012**, *29*, 449–457. [CrossRef]
30. Idir, R.; Cyr, M.; Tagnit-Hamou, A. Role of the nature of reaction products in the differing behaviours of fine glass powders and coarse glass aggregates used in concrete. *Mater. Struct.* **2012**, *46*, 233–243. [CrossRef]
31. Starinieri, V.; Illingworth, J.M.; Hughes, D. Use of supplementary aggregates in mortars produced using a novel lime drying technique. *Proc. Inst. Civ. Eng. Constr. Mater.* **2019**, *172*, 305–313. [CrossRef]
32. Matos, A.M.; Sousa-Coutinho, J. Durability of mortar using waste glass powder as cement replacement. *Constr. Build. Mater.* **2012**, *36*, 205–215. [CrossRef]
33. Papadakis, V.G.; Tsimas, S. Supplementary cementing materials in concrete Part I: Efficiency and design. *Cem. Concr. Res.* **2002**, *32*, 1525–1532. [CrossRef]
34. Degirmenci, N.; Yilmaz, A.; Cakir, O.A. Utilization of waste glass as sand replacement in cement mortar. *Indian J. Eng. Mater. Sci.* **2011**, *18*, 303–308.
35. Rajabipour, F.; Maraghechi, H.; Fischer, G. Investigating the Alkali-Silica Reaction of Recycled Glass Aggregates in Concrete Materials. *J. Mater. Civ. Eng.* **2010**, *22*, 1201–1208. [CrossRef]
36. Ali, E.E.; Al-Tersawy, S.H. Recycled glass as a partial replacement for fine aggregate in self compacting concrete. *Constr. Build. Mater.* **2012**, *35*, 785–791. [CrossRef]
37. UNI 11471:2013. *Beni Culturali—Valutazione Della Pozzolanicità di un Materiale—Metodo Chimico (Saggio di Pozzolanicità)*; Ente Nazionale Italiano di Unificazione: Milan, Italy, 2013.
38. Lee, H.; Hanif, A.; Usmand, M.; Sime, J.; Oh, H. Performance evaluation of concrete incorporating glass powder and glass sludge wastes as supplementary cementing material. *J. Clean. Prod.* **2018**, *170*, 683–693. [CrossRef]
39. Jani, Y.; Hogland, W. Waste glass in the production of cement and concrete—A review. *J. Environ. Chem. Eng.* **2014**, *2*, 1767–1775. [CrossRef]
40. UNI EN 1015-11. *Metodi di Prova per Malte per Opere Murarie—Parte 11: Determinazione Della Resistenza a Flessione e a Compressione Della Malta Indurita*; Ente Nazionale Italiano di Unificazione: Milan, Italy, 2007.
41. UNI EN 1015-18. *Metodi di Prova per Malte per Opere Murarie—Determinazione del Coefficiente di Assorbimento D'acqua per Capillarità Della Malta Indurita*; Ente Nazionale Italiano di Unificazione: Milan, Italy, 2004.

42. Moropoulou, A.; Bakolas, A.; Bisbikou, K. Investigation of the technology of historic mortars. *J. Cult. Herit.* **2000**, *1*, 45–58. [CrossRef]
43. Rizzo, G.; Megna, B. Characterization of hydraulic mortars by means of simultaneous thermal analysis. *J. Therm. Anal. Calorim.* **2008**, *92*, 173–178. [CrossRef]
44. Rizzo, G.; Ercoli, L.; Megna, B.; Parlapiano, M. Characterization of mortars from ancient and traditional water supply systems in Sicily. *J. Therm. Anal. Calorim.* **2008**, *92*, 323–330. [CrossRef]
45. Ismail, Z.Z.; Al-Hashmi, E.A. Recycling of waste glass as a partial replacement for fine aggregate in concrete. *Waste Manag.* **2009**, *29*, 655–659. [CrossRef]
46. Parghi, A.; Alam, M.S. Physical and mechanical properties of cementitious composites containing recycled glass powder (RGP) and styrene butadiene rubber (SBR). *Constr. Build. Mater.* **2016**, *104*, 34–43. [CrossRef]
47. Du, H.; Tan, K.H. Properties of high volume glass powder concrete. *Cem. Concr. Compos.* **2017**, *75*, 22–29. [CrossRef]
48. Kamali, M.; Ghahremaninezhad, A. Effect of glass powders on the mechanical and durability properties of cementitious materials. *Constr. Build. Mater.* **2015**, *98*, 407–416. [CrossRef]

recycling

Article

Performance Assessment of Reclaimed Asphalt Pavement (RAP) in Road Surface Mixtures

Vítor Antunes [1,2,*], José Neves [1] and Ana Cristina Freire [2]

1   CERIS, Department of Civil Engineering, Architecture and Geo-Resources, Instituto Superior Técnico Universidade de Lisboa, Av. Rovisco Pais, 1049-001 Lisbon, Portugal; jose.manuel.neves@tecnico.ulisboa.pt
2   LNEC, National Laboratory of Civil Engineering, Av. Do Brasil 101, 1700-066 Lisbon, Portugal; acfreire@lnec.pt
*   Correspondence: vitorfsantunes@tecnico.ulisboa.pt

Abstract: Considerable amounts of Reclaimed Asphalt Pavement (RAP) are produced every year, as the road network requires maintenance to ensure the safety and comfort of its users. RAP is a 100% recyclable material and a useful fit to be re-introduced into another cycle without downgrading its functionality. Despite the current knowledge about the benefits associated with RAP use, it is not yet largely applied in several countries. This paper aims to validate, on the basis of both short- and long-term mechanical behaviours, the application of a bituminous mixture with a high RAP incorporation rate (75%) in road pavement wearing courses. A crude tall oil rejuvenator was used. Both short- and long-term oven ageing procedures were employed to simulate the ageing that occurs during mixture production and in-service life, respectively. The tests for validating the RAP mixture as an alternative solution comprised stiffness, resistance to fatigue, permanent deformation, and determination of the water sensitivity. Furthermore, the RAP bitumen mobilisation degree was evaluated and a mixing protocol was established. In comparison with virgin bituminous mixtures, it was found that, in general, the high RAP mixtures presented similar or better behaviour. The ageing process had a hardening effect namely in terms of stiffness and resistance to permanent deformation, without significant effects on the resistance to fatigue and water damage.

Keywords: ageing; bituminous mixture; mechanical behaviour; RAP; rejuvenator

Citation: Antunes, V.; Neves, J.; Freire, A.C. Performance Assessment of Reclaimed Asphalt Pavement (RAP) in Road Surface Mixtures. *Recycling* **2021**, *6*, 32. https://doi.org/10.3390/recycling6020032

Academic Editor: Michele John

Received: 14 April 2021
Accepted: 10 May 2021
Published: 13 May 2021

Publisher's Note: MDPI stays neutral with regard to jurisdictional claims in published maps and institutional affiliations.

1. Introduction

Reclaimed Asphalt Pavement (RAP) (nomenclature adopted in this study), or just Reclaimed Asphalt (RA), as referred to in EN 13108-8, corresponds to removed and/or reprocessed pavement material containing bitumen and aggregates or material deriving from surplus production [1]. When properly crushed and screened, RAP consists of high-quality and well-graded aggregates coated by bituminous mastic [2]. This material can be processed and used as secondary raw material to replace virgin aggregates and bitumen, and as such it is highly valuable for recycling purposes. RAP must be characterized by the following:

- Bitumen content and, in the case of high rates, bitumen properties.
- Grading, bulk density and other consensus properties of aggregates.

From the circular economy perspective, the material should be recycled and incorporated in similar applications without downgrading its utility [3,4]. Hence, the requirements established for some aggregate properties are met by considering that the RAP will replace, either totally or partially, both the virgin aggregates and the bitumen necessary to produce the mixture for that course.

The milled RAP from a single project usually has consistent characteristics in terms of gradation, bitumen content, aggregate and bitumen properties. However, some differences can be observed if the road section of the project has undergone several repaving interventions. RAP from a single project only needs separation and stockpiling of RAP particles by fraction and, if necessary, crushing of the larger particles followed by screening and separation. The control of gradation and reduction of larger particles enables efficient drying and heating when RAP is used in hot or warm recycled mixtures, hence leading to easier breaking down of RAP agglomerations during the mixing process [5,6].

Road authorities have been looking for different approaches to optimizing RAP use. Some guidelines were introduced in the specifications, which pointed to the need to optimize the management of these materials. Recycling solutions in pavement maintenance and rehabilitation (M&R) promote the extended use of materials, which leads to meeting sustainability policies. RAP recycling has been facing several obstacles. Indeed, the strict specifications and legislations enforced in some countries illustrate the poor confidence in RAP recycling in bituminous mixtures (RAP mixtures) [6–8]. With rare exceptions, the maximum RAP incorporation rates vary between 10% and 50%. The maximum rates are more restrictive for wearing courses. This value generally varies worldwide between 0% and 20% in some cases, and in some European countries, high RAP incorporation rates are allowed on the condition that the mixes fulfil the requirements for dense-graded surface mixtures [3,6,8,9].

The wearing courses of road pavements have limited service life periods since frequent M&R operations are necessary on the part of road authorities to guarantee adequate safety and comfort conditions for road users. These yearly operations carried out on the road networks result in the production of huge quantities of RAP and involve high costs. Hence, RAP should be more and more envisaged as a secondary raw material and the application of high RAP mixtures should be promoted.

The quality and the production process of materials strongly affect the performance of the bituminous mixture. Those factors acquire greater relevance in RAP mixtures, namely in the case of high rates, as RAP is a composite material containing aggregates, bitumen and, occasionally, additives. A correct design allows determining the type and proportion of those materials in the mixture.

The mixing procedure has impacts on the homogeneity of the mixture. The homogeneity is more relevant as the RAP content increases. RAP mixtures present issues regarding RAP sizes, mixing methods and diffusion mechanisms. The degree of RAP bitumen mobilization and blending with the new virgin bitumen is essential to guarantee the greatest homogeneity of the mixture [10–12].

In high RAP mixtures, it is fundamental to evaluate and define a laboratory mixing protocol capable of approximately simulating the in-plant mixing process, without requiring major changes in the current in-plant process. This will ultimately present benefits because it will reduce the investment in upgrading the plants for controlling RAP mixtures [13]. Due to the significant percentage of bitumen in the RAP composition, it can be mobilized to act as a binding element rather than as just a "black rock". The mobilization degree of the RAP bitumen can result in economic gains, due to the reduced need for virgin bitumen, the most expensive component of bituminous mixtures [12,14].

In general, RAP recycling introduces some positive changes in the physical and mechanical behavior of bituminous mixtures. Some of them are related to the RAP bitumen resulting in an increased mixture stiffness, which eventually leads to the modification of resistances to fatigue, cracking (under low-temperature), and to permanent deformation [8,10,15–22]. Some important conclusions can be drawn from the literature:

- RAP should be heated to promote blending between the virgin bitumen and the RAP bitumen, otherwise, RAP tended to act like an aggregate ("black rock"); this allowed taking the greatest advantage of RAP [5,10].
- RAP particles had a coarser gradation than the aggregates present in these particles; in this way, RAP aggregates gradation needed to be determined [3,5,7].
- RAP particles failure to completely break down and blend in with the virgin materials may cause an overall coarser mixture and consequently lead to more voids in the mineral aggregate (VMA) than expected [10,23,24].

- RAP heating time allowed the breaking down of particles during the mixing process and, consequently, a highest rate of blending occurred between the RAP and the new materials [10,23,24].
- The increase in RAP content led to an increased variability of the results obtained in laboratory studies; this was mainly related to RAP variability and mixing procedures [5,10,25].
- RAP storage in fractions allowed a greater control over the mixture homogeneity, which led to less variability in laboratory results [5,10,25].
- RAP mixtures that reveal a higher stiffness were usually likely to present shorter fatigue life and higher susceptibility to thermal cracking [10,15,16,18,19].
- The in-service performance of recycled mixtures was in accordance with laboratory results. In terms of long-term performance, the RAP mixture presented less than or equal life to the virgin mixtures when compared against performance indicators [17,26–28].
- The recycling agent or rejuvenator allowed to partially regenerate RAP bitumen properties, which led to an easier blending between both the virgin and the RAP bitumen and the materials and, consequently, to lower stiffness and resistance to permanent deformation; as well as to higher fatigue and greater thermal cracking resistance [12,21,22,29–35].

RAP recycling in bituminous mixtures is not a novelty. However, this subject continues to generate renewed interest, due to the lack of confidence in this type of solution by the road authorities, namely in applications with high incorporation rates. Despite intensive investigation, the aspects that can contribute to obtaining homogeneous and good performance RAP bituminous mixtures have not been evaluated together. Hence, this paper aims to:

- Assess the RAP bitumen mobilization degree by a qualitative methodology.
- Evaluate the effect of RAP fractioning and mixing condition on mixture homogeneity.
- Determine the mechanical behavior of RAP mixtures with a high incorporation rate.
- Assess the long-term behavior of RAP mixtures by evaluating the impact of ageing on mechanical performance.

In sum, the main objective of this paper is to validate high RAP mixtures for wearing courses on basis of their long-term mechanical behavior. Hence, the previous aspects were evaluated together, with a view to contribute to a methodology that will lead to an increased confidence in RAP solutions. A comparison was established between high RAP mixtures (75% per total mass weight) and virgin bituminous mixtures. A dense-graded bituminous mixture for wearing courses with a maximum aggregate size of 14 mm was taken as a reference for all the mixtures.

2. Materials

2.1. RAP

The RAP was obtained from the rehabilitation of a motorway consisting of the replacement of the wearing course. The RAP was collected and its grading was characterized according to EN 933-1. RAP was fractionated in the laboratory and stocked in 3 fractions from the finest to the coarsest fraction: fine fraction 0/4.75 mm (RAP F), medium fraction 4.75/12.5 mm (RAP M) and coarse fraction 12.5/19 mm (RAP C).

2.2. Rejuvenator

The rejuvenator agent was a chemical by-product derived from Crude Tall Oil (CTO), a renewable raw material used in the paper production industry and consisting of a mixture of both fatty and rosin acids. This rejuvenator can act both as a softener and as a dispersant/compatibilizer/remobilizer of the RAP bitumen [36]. This is a rejuvenator that has already shown good results in previous studies [22,29,37].

2.3. Aggregates

This study involved fractions of basalt aggregates (0/4 mm, 4/12 mm, and 10/16 mm), limestone aggregates (0/4 mm) and commercial filler.

2.4. Bitumen

The bitumen used for producing the bituminous mixtures was a 35/50 nominal penetration neat bitumen, in accordance with EN 12591. It had a penetration of 44×10^{-1} mm, a softening point of 54.6 °C and a penetration index of -0.40.

3. Experimental Procedures

3.1. Bituminous Mixture Characterization

The studied dense graded bituminous mixture for wearing courses had a maximum aggregate size of 14 mm and was produced with a neat bitumen 35/50 (AC14 surf 35/50). Two compositions were studied: one without RAP (0RAP and 0RAPA); and another with a high incorporation RAP rate, 75% per total mass of mixture (75RAP0 and 75RAP0A). Both unaged (0RAP and 75RAP0) and aged conditions (0RAPA and 75RAP0A) were studied on the two compositions.

Bituminous mixture design was based on EN 12697-34, as a first approach to determine the optimum bitumen content. A set of cylindrical specimens were produced with different total bitumen contents (including RAP bitumen), 4.0%, 4.5%, 5.0% and 5.5%. The bituminous mixture composition was determined on basis of the volumetric and global performance behaviors.

Figure 1 summarizes the bituminous mixture design, from the selection of materials until the performance assessment of the bituminous mixtures, as well as the formulation and ageing processes.

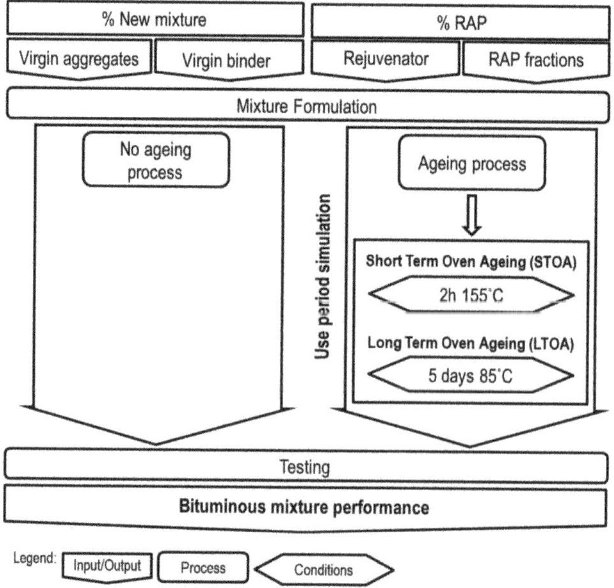

Figure 1. Bituminous mixture testing protocol.

Prismatic specimens were produced and compacted using the roller compactor according to EN 12697-33. Their larger dimensions allowed a greater homogeneity of the mixture. As the RAP mixtures consist of different fractions of RAP and virgin aggregates, small amounts of each fraction are necessary for cylindrical specimens, namely for RAP C, which may imply a reduced number of particles. Taking into consideration the heterogeneity of RAP particles, different quantities of coarser aggregates and RAP bitumen could be introduced in each specimen. These differences were less significant in the case of larger specimens. Furthermore, the use of prismatic specimens allowed similar ageing, unlike what happens when different specimen sizes are used, which biased the conclusions in terms of the long-term behavior.

3.2. RAP Characterization

The characteristics of RAP fractions were determined by the following tests:

- Aggregate particle size distribution according to EN 933-1 and EN 12697-2.
- Bitumen content according to EN 12697-1.
- Bitumen needle penetration and softening point according to EN 1426 and EN 1427, respectively.

To evaluate the RAP bitumen characteristics, the bitumen of the medium and coarse RAP fractions were recovered using the rotary evaporator method according to EN 12697-3. The bitumen from RAP F was not evaluated due to the high content of fine material. However, as RAP had the same origin it was assumed that the bitumen present in all fractions had the same characteristics. The recovered RAP bitumen was characterized by the needle penetration and the softening point.

3.3. Rejuvenator Dosage

Three percentages of rejuvenator (3.0%, 4.5% and 7.0%) were used to determine the optimum rejuvenator dosage. The tests were performed on recovered RAP bitumen samples and they were treated with rejuvenator in accordance with the following blending protocol:

1. The RAP bitumen was placed in an oven at 135 °C for 30 to 60 min, the time required for it to be fluid.
2. Afterwards, the RAP bitumen was mixed for homogenization and removal of any bubble.
3. The mass of RAP bitumen was determined and the container was placed on a hot plate at 100 to 135 °C.
4. The dosage of rejuvenator was added to RAP bitumen and hand-blended using a glass rod for about 30 s.
5. The container with the blend (RAP bitumen + rejuvenator) was placed in the oven at 135 °C for 10 min and the blend was mixed for 30 s at 5 min intervals.

After the treatment process, RAP bitumen samples were tested to penetration and softening point values. The literature indicates that the optimum percentage of rejuvenator by weight of RAP bitumen is the one that allows complying with the properties of the virgin bitumen percentage used in the RAP mixture [38]. In this study, the reference bituminous mixture is an AC14 surf 35/50, resulting in property limits for the treated bitumen similar to the 35/50 neat bitumen; the penetration varied between 35 and 50 mm at 25 °C and the softening point between 50 °C and 58 °C.

3.4. Mixing Conditions

To qualitatively assess how mixing conditions and the use of rejuvenator affect the mobilization of RAP bitumen and the homogeneity of the mixture, a set of 4 cylindrical specimens were produced using a blue pigmented bitumen. This option allows obtaining a visual contrast between the black from RAP bitumen and the blue from the virgin bitumen, which makes it possible to easily identify the RAP bitumen film coating the RAP aggregates. The pigment dosage was 2% per weight of the mixture, to obtain a strong color, and the virgin bitumen dosage was 4%, by assuming that no RAP bitumen would be mobilized. The addition was performed at room temperature because the bitumen and the pigment were provided in pellets. It was not necessary to perform to any corrections to the granulometric curve as the pigment was supplied in pellets and it melted down into the bitumen.

The specimens were produced considering the aspects as follows: the theoretical grading curve defined for the bituminous mixtures; the conditions in terms of temperature of materials; and the use of rejuvenator for the RAP mixtures. The following types of bituminous mixtures and specimens were produced:

- 0RAPH: produced with virgin aggregates heated in a ventilated oven at 165 °C for 4 h.
- 75RAPH: produced with 75% RAP and 25% virgin aggregates heated in a ventilated oven at 165 °C for 4 h together.
- 75RAPL: produced with 75% RAP and 25% virgin aggregates heated in a ventilated oven at 130 °C for 2.5 h and 205 °C for 4 h, respectively.
- 75RAPLR: produced with 75% RAP and 25% virgin aggregates heated in a ventilated oven at 130 °C for 2.5 h and 205 °C for 4 h, respectively, and using 4.5% of rejuvenator by mass of RAP bitumen.

The aggregates of the 0RAPH mixture were heated at 165 °C and mixed for 3 min with the virgin bitumen and the pigment pellets, added at room temperature. The same procedure was followed for 75RAPH, with the RAP being heated together with the virgin aggregates, which were both mixed with the virgin bitumen and the pigment pellets. The procedure presented a slight difference in the case of 75RAPL. The over-heated aggregate (205 °C) was mixed with the RAP for 30 s to allow the RAP to heat to the mixture temperature. Afterwards, the bitumen and the pigment pellets were added to the bowl and mixed for 150 s. Regarding the 75RAPLR, the process was more complex. In the case of rejuvenator addition combined with a low RAP temperature, the RAP was mixed with the rejuvenator for 30 s. Then, the over-heated aggregate was added and mixed for 60 s. Finally, the bitumen and the pigment pellets were added and mixed for 90 s. All the mixtures were compacted using the impact compactor.

3.5. Bituminous Mixture Production and Ageing

The bituminous mixture without RAP was produced according to EN 12697-35. For the RAP mixture, the standard protocol was modified to include a previous treatment of RAP with the rejuvenator. The protocol was intended to simulate the in-plant mixture production with pre-heating of RAP and the treatment with rejuvenator. The heating of RAP at lower temperatures avoids the hardening of the RAP bitumen and the consequent blue smokes [3,5,6,13]. In the case of high RAP mixtures, these changes in current methodology have substantial impacts. Thus, the RAP fractions were heated at 130 °C for 2.5 h. To achieve the mixture temperature for a 35/50 nominal penetration, the aggregate was superheated at 205 °C for 4 h, 40 °C more than the recommended mixing temperature. The virgin bitumen was heated at 165 °C for 3 h, as defined in the specification. The mixture procedure for the RAP mixture was the following:

- In the mixing bowl, the rejuvenator was spread over the heated RAP and mixed for 30 s.
- The over-heated aggregates were added to the mixing bowl and mixed for further 60 s.
- The virgin bitumen was added and mixed for 90 s.
- The mixture was poured into the mold and compacted.

To assess the performance of mixtures at a later stage of its service, the ageing process was simulated in the laboratory. The ageing methodology consisted of both short- (STOA) and long-term oven ageing (LTOA) according to AASHTO R30. Two sets of unaged and aged specimens were produced.

3.6. Mechanical Tests

The mixture performance was assessed in terms of mechanical properties: stiffness, resistance to fatigue, determination of the water sensitivity, and wheel-tracking (permanent deformation). Table 1 presents the test standards and conditions.

The specimens were compacted in prismatic molds (roller compactor method) and, afterwards, they were either sawed, to obtain small prismatic specimens for stiffness and fatigue resistance tests, or cored to obtain cylindrical specimens for the water sensitivity test.

Table 1. Mechanical tests.

Scope (Standard)	Conditions
Stiffness (EN12697-26)	Method: Four-point bending test (4PBT) Temperature: 20 °C Frequencies: 0.1, 0.2, 0.3, 0.4, 0.5, 0.6, 0.7, 0.8, 0.9, 1, 2, 3, 4, 5, 6, 7, 8, 9, 10, 15, 20, 25, 30, 1 and 0.1 Hz Strain: 50 µm/m
Resistance to fatigue (EN 12697-24)	Method: 4PBT Temperature: 20 °C Frequency: 10 Hz Strains: 200, 300 and 400 µm Failure criteria: complex stiffness modulus reaching half of its initial value
Determination of the water sensitivity (EN 12697-12)	Method A: Indirect tensile strength Temperature: 15 °C Specimen conditioning: 72 h (dry and wet conditions)
Wheel tracking (EN 12697-22)	Method: Small size device; Procedure B in air Temperature: 60 °C Specimen conditioning: 6 h at 60 °C

4. Results and Discussion

4.1. RAP Properties

Figure 2 shows the grading curves of the RAP sample representative of both the stockpile and the 3 laboratory fractions. For the latter, both the bitumen content and the recovered RAP aggregate grading were determined. In terms of bitumen content, RAP F presented the highest value, 6.4%, followed by RAP C with 3.4% and RAP M with 3.3%.

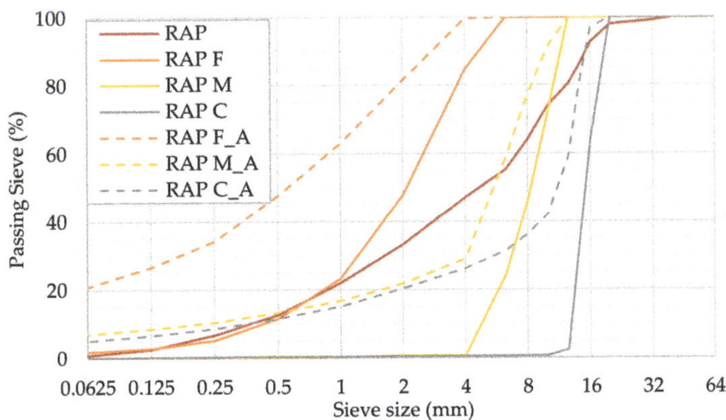

Figure 2. Grading curves of RAP.

After the determination of the bitumen content, the grading of RAP aggregates from each fraction was assessed. Figure 2 shows the grading curves of RAP, RAP fractions and RAP aggregates (designation followed by "A"). Despite fractioning, the RAP aggregates had a finer gradation than the fraction limits. This result was expected because the fraction was performed on RAP particles containing agglomerations of aggregates covered and bonded by mastic (RAP bitumen and fine material). Therefore, RAP M and RAP C showed a considerable amount of material passing the 0.063 mm sieve. Hence, the grading curves of the aggregates recovered from the RAP fractions were used for the RAP mixture formulation.

Table 2 presents the results of the penetration determined at 25 °C for RAP M and RAP C bitumen. RAP F presented a high fine material percentage during the bitumen content determination, which may influence the penetration of the bitumen recovered from this fraction; if the fraction was tested, it would be expected to lead to values similar to a bituminous mastic rather than to a recovered bitumen. For this reason, the results for this fraction will not be representative of the real ageing and the consequent hardening of the bitumen during the lifespan of the pavement course. The two coarsest fractions showed similar values of penetration, which demonstrates that the RAP ageing was homogeneous; the RAP bitumen presented an average penetration value of 19.5×10^{-1} mm.

Table 2. Penetration, softening point and penetration index of RAP bitumen and rejuvenated bitumen.

Sample	Softening Point (°C)	Penetration 25 °C ($\times 10^{-1}$ mm)	Penetration Index
RAP M	68	18	0.22
RAP C	67	21	0.38
RA + 3.0% CTO	63	30	0.33
RA + 4.5% CTO	61	36	0.43
RA + 7.0% CTO	53	61	0.09
Neat 35/50	50–58	35–50	—

The softening point results are also shown in Table 2. The results of the aged bitumen recovered from RAP M and RAP C were evaluated. The results are similar for all samples (average value of 67.5 °C) which also confirms that the RAP ageing was homogenous.

As the RAP bitumen from laboratory fractions had the same origin, they presented similar softening point and penetration values. These results confirm the homogeneity of the ageing process. Thus, a single rejuvenator percentage was determined for all RAP fractions.

4.2. Optimum Rejuvenator Dosage

Three percentages of rejuvenator by weight of RAP bitumen were tested: 3.0%, 4.5% and 7.0%. The results are shown in Table 2. The 3.0% CTO did not fit with the values of a 35/50 nominal penetration bitumen. For the highest percentage, the penetration value obtained was 22% higher than the maximum limit. The 4.5% CTO allowed obtaining a penetration value that fits the limit. However, the softening point was 5% higher than the maximum limit. The difference in the softening point was not critical and 4.5% of rejuvenator was considered to be the optimum value that allows minimizing costs.

4.3. Mobilization of RAP Bitumen

Figure 3 shows that the RAP specimens presented a smoother surface than the 0RAPH. The excess of bitumen detected in the rim of the mold is an indicator of excessive bitumen content, leading to conclude that more RAP bitumen might have been mobilized than expected.

The tops of the specimens were trimmed and afterwards cut to obtain four semi-circular samples. The cut specimens' surfaces were observed to evaluate the homogeneity of the mixtures. Figure 4 presents a comparison between mixtures with and without RAP for different mixing conditions. It was possible to observe that, at a macroscale, no RAP bitumen film surrounded the aggregates and that the RAP bitumen had darkened the mixture, which evidenced that most of the RAP bitumen was mobilized. Additionally, the mixture seemed to be homogeneous, presenting a small bluer area for the mixture 75RAPLR not darkened by the RAP bitumen. This area may result from a pellet of blue pigment that had not been totally blended into the mixture. This can be due to the shorter mixing time after the addition of the pigmented bitumen.

Figure 3. Blue specimen with an excess of bitumen.

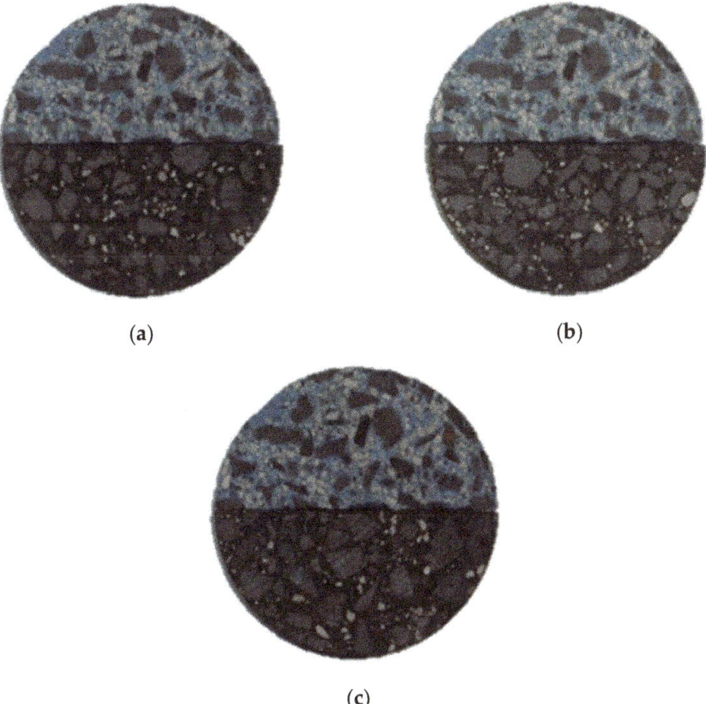

Figure 4. Blue specimens half-cut view of bituminous mixtures without RAP and RAP mixtures:
(**a**) 0RAPH-75RAPH; (**b**) 0RAPH-75RAPL; (**c**) 0RAPH-75RAPLR.

The use of low temperatures to heat the RAP was able to mobilize the RAP bitumen and hence avoid blue smoke emissions and additional ageing. Furthermore, the use of under-heated RAP combined with over-heated aggregates led to achieving the required mixing temperatures.

This qualitative analysis demonstrated that there was full or near to full mobilization of the RAP bitumen. Only a micro film of RAP bitumen surrounding RAP aggregates could probably be seen if observed in the microscope [11]. In this way, full mobilization of RAP bitumen was assumed in this study and the mixing protocol adopted made it possible to take the most advantage of the RAP bitumen.

4.4. Composition and Volumetric Properties

Table 3 shows the composition and properties of the mixtures. The same virgin aggregate fractions and virgin bitumen were used in the mixtures with and without RAP. In terms of total bitumen, both mixtures satisfied the minimum content (4%) specified for this type of mixture and application. Different total bitumen contents were defined for the mixtures, as the best behavior was verified with different bitumen contents. As RAP aggregates were coated by bitumen the adsorption and absorption phenomena had already occurred; the bituminous mixture presenting a lower total bitumen content due to a decreased need for bitumen. Moreover, as the finest part of the RAP mixture came from the RAP fractions, the filler was already coated by bitumen and this material, with a higher specific surface, did not mobilize as much bitumen as the virgin filler.

Table 3. Composition and properties of the bituminous mixtures.

Constituents		0RAP | 0RAPA	75RAP0 | 75RAP0A	Required Limits
Bitumen	Virgin: 35/50 paving grade bitumen (%)	5	1.5	-
	RAP (%)	0	2.8	-
	Total content %	5	4.3	4
Natural aggregates (%)	Fraction 10/16: Basalt	24	7	-
	Fraction 4/12: Basalt	33	4	-
	Fraction 0/4: Basalt	11	8	-
	Fraction 0/4: Limestone	29	6	-
	Commercial filler	3	0	-
RAP (%)	RAP F	0	8	-
	RAP M	0	44	-
	RAP C	0	23	-
Additives (% of RAP bitumen)	CTO	0	4.5	-
VMA (%)		15.5	14.2	Min 14
VFB (%)		79.5	76.1	-
Porosity (%)		3.2	3.4	3–5
Marshall stability (kN)		14.8	21	7.5–15/21
Marshall flow (mm)		3.0	3.2	2 4
Marshall coefficient (kN/mm)		4.9	6.6	Min 3
ρ_{bssd} (Mg/m^3)		2.56	2.61	-
ρ_{mv} (Mg/m^3)		2.64	2.69	-
F/B ()		1.4	1.8	-

The grading curves of RAP fraction aggregates were used to establish the final grading curve of the RAP mixture. It was assumed that the RAP bitumen was fully mobilized and blended with the virgin bitumen. Thus, the full 2.8% of RAP bitumen percentage was considered in the mixture design. A percentage of 4.5% of rejuvenator was added to the mixture.

Figure 5 presents the aggregate grading curves of bituminous mixtures, within the required limits [39,40]. The RAP grading curve presents a higher content of fine material, which can be explained by the high RAP content.

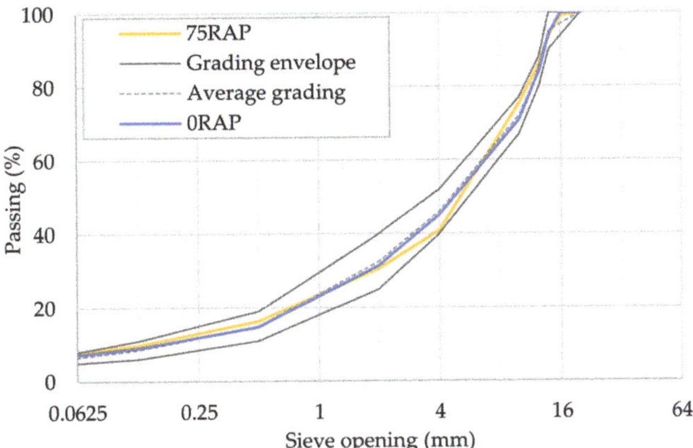

Figure 5. Aggregate grading curves.

Marshall tests according to EN 12697-34 were carried out to measure the stability and flow properties. The following volumetric properties of the test specimens were also evaluated: maximum specific density (ρ_{mb}), bulk specific density (ρ_{bssd}), void content (V), voids in mineral aggregates (VMA), and voids filled with bitumen (VFB). Table 3 presents the average values obtained for those properties.

The total bitumen content of the RAP mixture was 0.7% lower than the one of virgin mixture (0RAP), which had a direct influence on V and VFB that were higher and lower, respectively. Despite the higher percentage of fine material of the RAP mixture when compared with the virgin mixture, the V of the mixture was not strongly affected. The fractioning and control of RAP allowed a greater control over the final RAP mixture characteristics.

The Marshall properties were only determined for the unaged bituminous mixtures. The unaged RAP mixture (75RAP0) presented higher stability, flow and coefficient when compared with 0RAP. In general, the values obtained complied with the required limits [39,40]. However, 75RAP0 presented a stability value equal to the maximum limit. Since the origin of the RAP aggregates is unknown, such a higher limit was assumed.

4.5. Stiffness Behaviour

Figure 6 shows the stiffness modulus (E*) and the phase angle of the bituminous mixtures. Unaged RAP mixture (75RAP0) presented an increase of 35% in the stiffness modulus for all tested frequencies and an increase of 18% for frequencies between 1 HZ to 30 Hz, when compared with the unaged virgin mixture (0RAP). These results are in accordance with the high values of Marshall stability. The increase in the test frequency led to a reduction in the stiffness difference due to the approximation of the glassy modulus verified at higher frequencies and lower temperatures.

The ageing procedure applied had a strong effect on the increase of the stiffness modulus. Similarly to the comparison between unaged virgin mixture (0RAP) and unaged RAP mixture (75RAP0), the difference verified for each frequency decreased with the increase in frequency. Average increases of 13% and 32% were observed when comparing the unaged virgin mixture with the aged one and the unaged RAP mixture with the aged one, respectively. The RAP mixture was more susceptible to ageing.

Regarding the phase angle, the introduction of RAP led to an average decrease of 13%. This effect is related to the RAP bitumen that contributes to the hardening of the bituminous mastic.

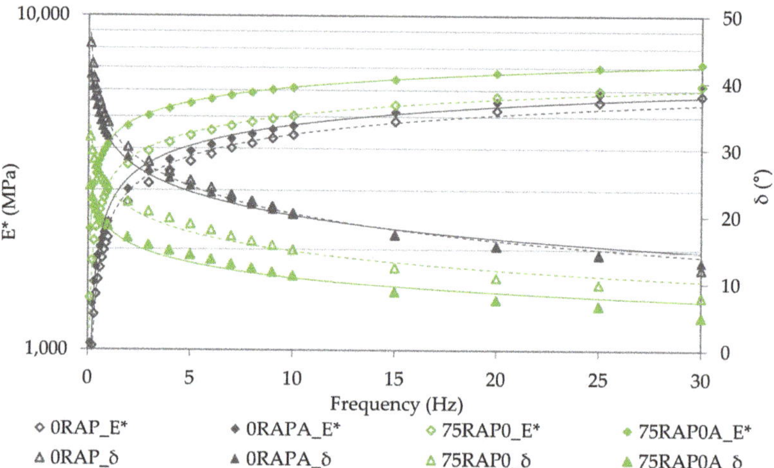

Figure 6. (**left**) Stiffness modulus; (**right**) Phase angle. E* is the stiffness modulus

The ageing process contributes to the hardening of both bituminous mastics due to the alterations in the chemical and physical properties, which lead to losses of bitumen volatile parts and to alterations in the viscoelastic behavior. In consequence, a change occurred in the mechanical response of mixtures under load conditions, which corresponded to a decrease in the phase angle. The ageing contributed to average decreases of 4% for the virgin mixture and of 38% for the RAP mixture.

Figure 7 presents the Cole-Cole and the Black diagrams of the rheological properties of the bituminous mixtures. The Black diagram expresses the correlation between the stiffness modulus and the phase angle. Figure 7b shows a phase angle reduction from the unaged virgin mixture (0RAP) to the aged RAP mixture (75RAP0A), which is associated with the differences in bitumen type and ageing. The new bitumen and the rejuvenator of the RAP mixture contributed to reduce bitumen viscosity. Some differences observed in terms of stiffness modulus and phase angle may also be associated with the differences found in the mineral skeleton of bituminous mixtures. The RAP mixture presented a higher content of fine material, which led to a different aggregate configuration and to contacts between particles. Furthermore, the ageing process contributed to bitumen hardening and hence to a loss of viscosity of the bitumen mastic.

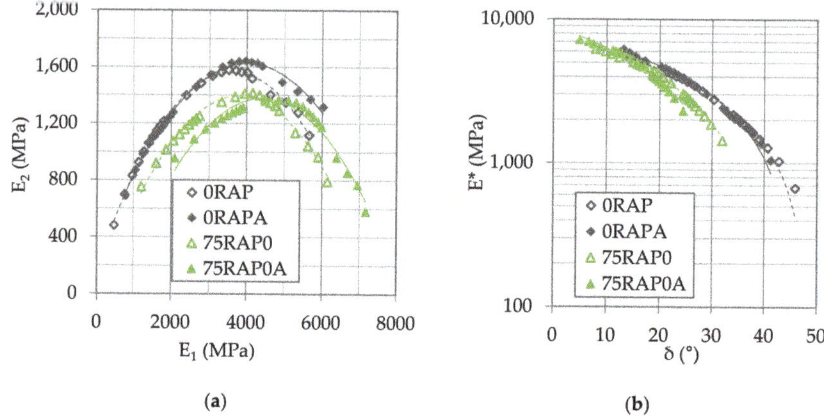

Figure 7. (**a**) Cole-Cole diagram; (**b**) Black diagram. E* is the stiffness modulus.

The ageing led to a decrease in the phase angle and to an increase in stiffness, which was due to the contribution of the elastic part of the mixture.

Figure 7a confirmed the previous considerations. The 0RAP showed the highest values for both the elastic (E_1) and the viscous (E_2) components of the complex stiffness modulus (E^*), for the same load and temperature conditions. RAP recycling led to the decrease of E1 and E2 when compared with the unaged virgin mixture (0RAP). The ageing protocol had a considerable effect on the rheological properties of mixtures, which is demonstrated by the increase in E_1 of the aged mixtures with the reduction in E_2, for the same load, temperature, and frequency conditions.

4.6. Resistance to Fatigue

Table 4 presents the coefficients a and b of the fatigue laws ($\varepsilon = a \times N^b$) obtained for three levels of strains. Figure 8 presents the extensions (ε_4, ε_5 and ε_6) to failure for each mixture as a function of three levels of applied loads: 10^4, 10^5 and 10^6, respectively. The ε_6 parameter is commonly used as a fatigue performance limit for the bituminous mixtures. The bituminous mixtures revealed similar behaviour based on this parameter. However, the unaged RAP mixture (75RAP0) revealed in general higher values for the three levels of failure strains.

Table 4. Experimental coefficients of fatigue laws.

Bituminous Mixtures	a	b	R^2
0RAP	3151.28	−0.21	0.96
0RAPA	1831.03	−0.17	0.91
75RAP0	3408.09	−0.22	0.89
75RAP0A	1937.78	−0.18	0.89

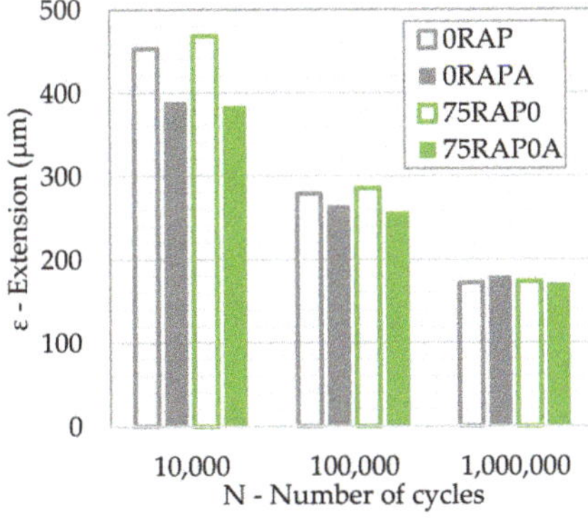

Figure 8. Extension to failure as a function of the number of loads 10^4, 10^5 and 10^6.

Contrarily to the expectations, the increase in stiffness resulted from RAP having no influence on the fatigue performance. It can be concluded that RAP recycling, even with high rates, did not compromise fatigue resistance.

The ageing process led to a reduction in the fatigue resistance expressed by the average decreases of 5% for the virgin mixture and 10% for the RAP mixture. In the case of the aged virgin mixture (0RAPA) a slight increase was observed for ε_6 4.7%. The reductions in fatigue resistance were due to the hardening of bituminous mastics.

4.7. Resistance to Moisture Damage

Figure 9 shows the water sensitivity evaluated by the indirect tensile strength ratio (ITSR), as well as both dry (ITSd) and wet (ITSw) indirect resistances. All the bituminous mixtures revealed good performance, achieving an ITSR value higher than 80% (ITSR80), the minimum required limit [41]. The mixtures presented low sensitivity to moisture damage and they withstood the harmful action of water. The unaged RAP mixture (75RAP0) presented the best results (ITSR = 96%) followed by the unaged (0RAP) and aged (0RAPA) virgin mixtures. Additionally, the introduction of RAP increased by 20% both the ITSw and the ITSd. The unaged RAP mixture (75RAP0) showed good moisture resistance, associated with higher indirect resistances, in accordance with the higher values of stiffness and Marshall stability.

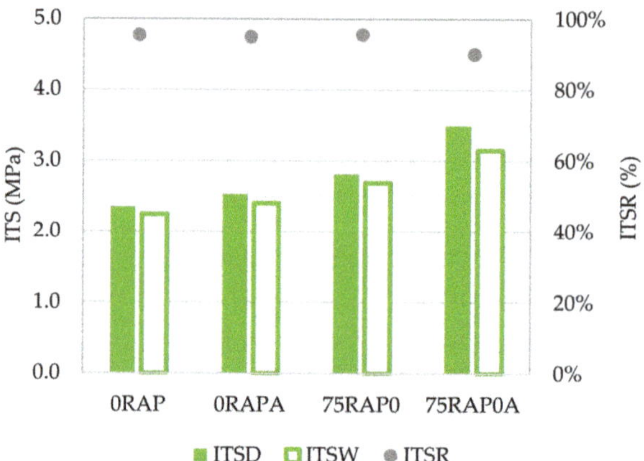

Figure 9. Resistances to moisture damage.

The ageing did not affect the virgin mixture and decreased the ITSR of the RAP mixture (6%), reaching a value of 90%, and showing good resistance to moisture damage. Nonetheless, this lower value in terms of ITRS is correlated with higher values in terms of indirect tensile resistances due to ageing: aged virgin mixture (0RAPA) presented increases of 8% and 7% and aged RAP mixture (75RAP0A) showed increases of 24% and 17% for ITSd and ITSw, respectively.

4.8. Resistance to Permanent Deformation

Figure 10a presents the deformation curves obtained from Wheel-tracking tests (WT). In terms of mean rut depth value (RD_{AIR}). The mixtures can be ranked from the highest to the lowest value as unaged (0RAP) and aged (0RAPA) virgin mixtures followed by unaged (75RAP0) and aged (75RAP0A) RAP mixtures, with 5.9 mm, 3.4 mm, 2.0 mm and 1.6 mm, respectively. The parameters as follows were determined from the curves obtained: slope in air (WTS_{AIR}) and mean proportional rut depth (PRD_{AIR}) (Figure 10b). Unaged RAP mixture (75RAP0) presented 66%, 90% and 66% reductions for RD_{AIR}, WTS_{AIR} and PRD_{AIR}, respectively, when compared with the unaged virgin mixture (0RAP). This behavior pointed out the clear contribution of RAP to the resistance to permanent deformation. This accrued resistance was due to the finer aggregate skeleton, which contributed to the increased number of contacts between particles, and to the presence of aged mastic from RAP.

The ageing procedure adopted proved to have a discernible effect on the permanent deformation resistance when the aged and unaged mixtures were compared. The aged virgin mixture (0RAPA) presented a decrease of 43%, 71%, and 42% and the aged RAP mixture (75RAP0A) showed a decrease of 20%, 3.7% and 19% for RD_{AIR}, WTS_{AIR} and PRD_{AIR}, respectively. These decreases for all variables are strongly related to the ageing of the bituminous mastic that occurred during the STOA and LTOA.

The chart in Figure 10a also shows that both RAP mixtures presented a tendency to a limit value, while the unaged virgin mixture (0RAP) tended to continue increasing the value of the maximum rut depth. Aged RAP mixture (75RAP0A) presents better results for the three parameters measured by the wheel tracking test.

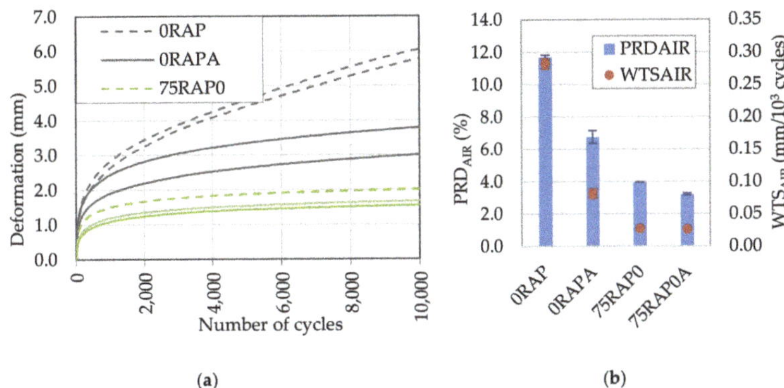

Figure 10. Permanent deformation performance: (**a**) wheel-tracking deformation curves; (**b**) WTS$_{AIR}$ and PRD$_{AIR}$.

5. Conclusions

The study presented in this paper investigated the possibility of incorporating high RAP rates in hot recycled bituminous mixtures for road wearing courses. The RAP was firstly fractionated into fine (RAP F—0/4.75 mm), medium (RAP M—4.75/12.5 mm) and coarse (RAP C—12.5/19 mm) fractions, according to the dimensions of the RAP particles. Two bituminous mixtures were studied: one virgin mixture (0RAP and 0RAPA) and one RAP mixture containing 75% RAP (75RAP0 and 75RAP0A) with rejuvenator, both unaged and aged to simulate the entire lifespan when in-service. All the bituminous mixtures fulfilled the properties required for a dense graded bituminous mixture with a maximum aggregate nominal size of 14 mm and a 35/50 nominal penetration bitumen, for road wearing courses.

The main findings of the study can be summarized as follows:

- Despite the unknown provenance and ageing of RAP, with just a mere fractioning and characterization of RAP bitumen and aggregate grading it was possible to formulate a high RAP mixture.
- The qualitative assessment of bitumen mobilization revealed that RAP bitumen can be fully or quasi fully mobilized when RAP is heated, allowing to take advantage of the RAP bitumen contribution as a binding element and not just as "black rock".
- The use of a rejuvenator derived from Crude Tall Oil (CTO) allowed recovering the RAP bitumen viscosity properties.
- Generally, the unaged RAP mixture had a better or similar mechanical performance when compared with the virgin mixture.
- Unaged RAP mixture presented a higher stiffness modulus when compared with the virgin one. The ageing protocol further increased the stiffness modulus.
- Despite the increment in resistance observed in the Marshall stability and in the stiffness modulus, the unaged RAP mixture presented a better behavior in terms of fatigue resistance. This can be related to the presence of a rejuvenator and the finer gradation of the bituminous mixture.
- The ageing did not affect the fatigue resistance of the RAP mixture, which demonstrated good long-term behavior.
- The introduction of RAP strongly affected the rutting resistance of the mixtures. This can be related to the lower total bitumen content in the mixture and the presence of RAP bitumen. The ageing increased the rutting resistance of mixtures. This effect was due to the hardening of the bituminous mastic.
- The unaged RAP mixture revealed the highest moisture resistance, which demonstrates that the use of high RAP rates did not affect this property. Even after the ageing process, the mixture still presented good behavior.

The performance of RAP mixtures showed good results even after ageing, which made it possible to establish that high RAP mixtures can present good long-term performances. Hence, provided that a proper evaluation is carried out, high RAP mixtures combined with a rejuvenator to take advantage of the RAP bitumen, are suitable alternatives to conventional hot bituminous mixtures. The fractioning and stockpiling of RAP, combined with a relatively easy characterization, allow a greater control over the homogeneity of RAP mixtures and, hence, over their behavior.

Further investigations must be performed regarding the RAP multi-recycling capacity and the bitumen blend (RAP bitumen + rejuvenator + virgin bitumen) behavior to understand how this material can be recycled several times to fulfil the European goals to increase circular economy. Moreover, different RAP and bitumen sources must be considered.

Author Contributions: Conceptualization, V.A.; methodology, V.A.; validation, V.A., J.N. and A.C.F.; formal analysis, V.A., A.C.F. and J.N.; investigation, V.A.; resources, V.A., A.C.F. and J.N.; data curation, V.A.; writing—original draft preparation, V.A.; writing—review and editing, V.A., J.N. and A.C.F.; supervision, A.C.F. and J.N.; funding acquisition, V.A. All authors have read and agreed to the published version of the manuscript.

Funding: This research was funded by Portuguese agency that supports science and technology ("Fundação para a Ciência e Tecnologia"), grant number SFRH/BD/114715/2016.

Data Availability Statement: Not applicable.

Acknowledgments: The authors gratefully acknowledge the Portuguese agency that supports science and technology ("Fundação para a Ciência e a Tecnologia") for the financial support provided through the grant SFRH/BD/114715/2016, and CERIS (Civil Engineering Research and Innovation for Sustainability).

Conflicts of Interest: The authors declare no conflict of interest.

References

1. CEN, EN 13108-8. *Bituminous Mixtures—Material Specifications—Part 8: Reclaimed Asphalt*; European Committee for Standardization: Brussels, Belgium, 2005.
2. Chesner, W.; Collins, R.; MacKay, M.; Emer, J. FHWA-RD-97-148. In *User Guidelines for Waste and Byproduct Materials in Pavement Construction*; Federal Highway Administration: Washington, DC, USA, 1998. Available online: https://www.fhwa.dot.gov/publications/research/infrastructure/pavements/97148/index.cfm#content (accessed on 14 January 2021).
3. EAPA. *Asphalt the 100% Recyclable Construction Product EAPA Position Paper*; European Asphalt Pavement Association: Brussels, Belgium, 2014.
4. FIEC, EBC, EFBWW, UIPI, Eurima, BUILD EUROPE, RICS, CEMBUREAU, GCP EUROPE, EDA, ECCE, EAPA, EUROPEAN CONCRETE PLATFORM, EUPAVE, EFCA, CONSTRUCTION 2050—Building Tomorrow's Europe Today. 2019. Available online: https://www.fiec.eu/download_file/force/539/510 (accessed on 15 March 2020).
5. West, R.C. *QIP 129—Best Practices for RAP and RAS Management*; Lanham: Prince George's County, MD, USA, 2015.
6. Antunes, V.; Freire, A.; Neves, J. A review on the effect of RAP recycling on bituminous mixtures properties and the viability of multi-recycling. *Constr. Build. Mater.* **2019**, *211*, 453–469. [CrossRef]
7. Zaumanis, M.; Mallick, R.B.; Frank, R. 100% recycled hot mix asphalt: A review and analysis. *Resour. Conserv. Recycl.* **2014**, *92*, 230–245. [CrossRef]
8. Zaumanis, M.; Arraigada, M.; Wyss, S.; Zeyer, K.; Cavalli, M.; Poulikakos, L. Performance-based design of 100% recycled hot-mix asphalt and validation using traffic load simulator. *J. Clean. Prod.* **2019**, *237*, 117679. [CrossRef]
9. Petho, L.; Denneman, E. *Maximising the Use of Reclaimed Asphalt Pavement in Asphalt Mix Design Field Validation*; Austroads Research Report AP-R517-16; Austroads: Sydney, Australia, 2016.
10. Nguyen, V.H. Effects of Laboratory Mixing Methods and RAP Materials on Performance of Hot Recycled Asphalt Mixtures. Ph.D. Thesis, University of Nottingham, Nottingham, UK, 2009. Available online: http://etheses.nottingham.ac.uk/863/ (accessed on 15 January 2021).
11. Cavalli, M.; Griffa, M.; Bressi, S.; Partl, M.; Tebaldi, G.; Poulikakos, L. Multiscale imaging and characterization of the effect of mixing temperature on asphalt concrete containing recycled components. *J. Microsc.* **2016**, *264*, 22–33. [CrossRef] [PubMed]
12. Zaumanis, M.; Mallick, R.B. Review of very high-content reclaimed asphalt use in plant-produced pavements: State of the art. *Int. J. Pavement Eng.* **2015**, *16*, 39–55. [CrossRef]
13. Zaumanis, M.; Mallick, R.B.; Frank, R. 100% Hot Mix Asphalt Recycling: Challenges and Benefits. *Transp. Res. Procedia* **2016**, *14*, 3493–3502. [CrossRef]

14. Zhang, J.; Sun, C.; Li, P.; Jiang, H.; Liang, M.; Yao, Z.; Zhang, X.; Airey, G. Effect of different viscous rejuvenators on chemical and mechanical behavior of aged and recovered bitumen from RAP. *Constr. Build. Mater.* **2020**, *239*, 117755. [CrossRef]
15. Daniel, J.; Lachance, A. Mechanistic and Volumetric Properties of Asphalt Mixtures with Recycled Asphalt Pavement. *Transp. Res. Rec. J. Transp. Res. Board* **2005**, *1929*, 28–36. [CrossRef]
16. Shu, X.; Huang, B.; Vukosavljevic, D. Laboratory evaluation of fatigue characteristics of recycled asphalt mixture. *Constr. Build. Mater.* **2008**, *22*, 1323–1330. [CrossRef]
17. Valdés, G.; Pérez-Jiménez, F.; Miró, R.; Martínez, A.; Botella, R. Experimental study of recycled asphalt mixtures with high percentages of reclaimed asphalt pavement (RAP). *Constr. Build. Mater.* **2011**, *25*, 1289–1297. [CrossRef]
18. Huang, B.; Zhang, Z.; Kingery, W.; Zuo, G. Fatigue Crack Characteristics of HMA Mixtures Containing RAP. In *Fifth International RILEM Conference on Reflective Cracking in Pavements*; RILEM Publications SARL: Limoges, France, 2004; pp. 631–638.
19. Izaks, R.; Haritonovs, V.; Klasa, I.; Zaumanis, M. Hot Mix Asphalt with High RAP Content. *Procedia Eng.* **2015**, *114*, 676–684. [CrossRef]
20. Zaumanis, M.; Mallick, R.B.; Frank, R. Evaluation of Rejuvenator's Effectiveness with Conventional Mix Testing for 100% Reclaimed Asphalt Pavement Mixtures. *Transp. Res. Rec. J. Transp. Res. Board* **2013**, *2370*, 17–25. [CrossRef]
21. Elkashef, M.; Williams, R.C. Improving fatigue and low temperature performance of 100% RAP mixtures using a soybean-derived rejuvenator. *Constr. Build. Mater.* **2017**, *151*, 345–352. [CrossRef]
22. Bocci, E.; Grilli, A.; Bocci, M.; Gomes, V. Recycling of high percentages of reclaimed asphalt using a bio-rejuvenator—A case study. In Proceedings of the 6th Eurasphalt & Eurobitume Congress, Prague, Czech Republic, 1–3 June 2016. [CrossRef]
23. Lee, T.-C.; Terrel, R.L.; Mahoney, J.P. Test for Efficiency of Mixing of Recycled Asphalt Paving Mixtures. *Transp. Res. Rec. J. Transp. Res. Board.* **1983**, 51–60.
24. Al-Qadi, I.L.; Carpenter, S.H.; Roberts, G.; Ozer, H.; Aurangzeb, Q.; Elseifi, M.; Trepanier, J. *Determination of Usable Residual Asphalt Binder in RAP*; Research Report ICT-09-031; Illinois Center for Transportation: Rantoul, IL, USA, 2009.
25. Al-Qadi, I.L.; Aurangzeb, Q.; Carpenter, S.H.; Pine, W.J.; Trepanier, J. *Impact of High RAP Content on Structural and Performance Properties of Asphalt Mixtures*; Research Report ICT-R27-37; Illinois Center for Transportation: Rantoul, IL, USA, 2012.
26. Wang, Y. The effects of using reclaimed asphalt pavements (RAP) on the long-term performance of asphalt concrete overlays. *Constr. Build. Mater.* **2016**, *120*, 335–348. [CrossRef]
27. Stroup-Gardiner, M. *Use of Reclaimed Asphalt Pavement and Recycled Asphalt Shingles in Asphalt Mixtures*; The National Academies Press: Washington, DC, USA, 2016. [CrossRef]
28. Gong, H.; Huang, B.; Shu, X. Field performance evaluation of asphalt mixtures containing high percentage of RAP using LTPP data. *Constr. Build. Mater.* **2018**, *176*, 118–128. [CrossRef]
29. Gomes, V.; Di Nolfo, M.; Vlachos, P.; Bocci, M. Two case studies with high levels of RA enabled by a rejuvenating agent. In Proceedings of the 6th Eurasphalt Eurobitume Congress, Prague, Czech Republic, 1–3 June 2016.
30. Porot, L.; Di Nolfo, M.; Polastro, E.; Tulcinsky, S. Life cycle evaluation for reusing Reclaimed Asphalt with a bio-rejuvenating agent. In Proceedings of the 6th Eurasphalt Eurobitume Congress, Prague, Czech Republic, 1–3 June 2016; pp. 1–8. [CrossRef]
31. Hagos, E.; Shirazi, M.; Van De Wall, A. The development of 100% RAP asphalt mixture with the use of innovative rejuvenator. In Proceedings of the 6th Eurasphalt Eurobitume Congress, Prague, Czech Republic, 1–3 June 2016. [CrossRef]
32. Mogawer, W.S.; Booshehrian, A.; Vahidi, S.; Austerman, A.J. Evaluating the effect of rejuvenators on the degree of blending and performance of high RAP, RAS, and RAP/RAS mixtures. *Road Mater. Pavement Des.* **2013**, *14*, 193–213. [CrossRef]
33. Kowalski, K.J.; Król, J.B.; Radziszewski, P.; Piłat, J.; Sarnowski, M. New concept of sustainable road structure with RAP binder course using. In Proceedings of the 6th Eurasphalt Eurobitume Congress, Prague, Czech Republic, 1–3 June 2016. [CrossRef]
34. Mangiafico, S.; Sauzéat, C.; Di Benedetto, H.; Pouget, S.; Olard, F.; Planque, L. Complex modulus and fatigue performances of bituminous mixtures with reclaimed asphalt pavement and a recycling agent of vegetable origin. *Road Mater. Pavement Des.* **2016**, *18*, 315–330. [CrossRef]
35. Zhang, J.; Guo, C.; Chen, T.; Zhang, W.; Yao, K.; Fan, C.; Liang, M.; Guo, C.; Yao, Z. Evaluation on the mechanical performance of recycled asphalt mixtures incorporated with high percentage of RAP and self-developed rejuvenators. *Constr. Build. Mater.* **2021**, *269*, 121337. [CrossRef]
36. De Bock, L.; Piérard, N.; Vansteenkiste, S.; Vanelstraete, A. *Categorisation and Analysis of Rejuvenators for Asphalt Recycling—Dossier 21*; Belgian Road Research Centre: Brussels, Belgium, 2020.
37. Tran, N.; Taylor, A.; Turner, P.; Holmes, C.; Porot, L. Effect of rejuvenator on performance characteristics of high RAP mixture. *Road Mater. Pavement Des.* **2016**, *18*, 183–208. [CrossRef]
38. EAPA. *Recommendations for the Use Rejuvenators in Hot and Warm Asphalt Production*; European Asphalt Pavement Association: Brussels, Belgium, 2018.
39. EP. 14.03—Materials. In *Construction Specifications Book*; Infraestruturas de Portugal: Almada, Portugal, 2012. (In Portuguese)
40. EP. 15.03—Constructive Methods. In *Construction Specifications Book*; Infraestruturas de Portugal: Almada, Portugal, 2014. (In Portuguese)
41. Nikolaides, A. *Highway Engineering: Pavements, Materials and Control of Quality*; CRC Press—Taylor & Francis Group: Boca Raton, FL, USA, 2015.

Article

Mechanical Properties of Concrete with Recycled Concrete Aggregate and Fly Ash

Ihab Katar, Yasser Ibrahim *, Mohammad Abdul Malik and Shabir Hussain Khahro

Engineering Management Department, Prince Sultan University, Riyadh 11586, Saudi Arabia;
ikatar@psu.edu.sa (I.K.); abdulmalik@psu.edu.sa (M.A.M.); shkhahro@psu.edu.sa (S.H.K.)
* Correspondence: ymansour@psu.edu.sa; Tel.: +966-553470474

Abstract: Recycled concrete aggregate (RCA) collected from the demolition of old reinforced concrete structures can be reused to prepare structural and non-structural concrete, thereby protecting the environment by preserving natural resources. This study explores RCA's use, collected from the crushed concrete of different building projects in Riyadh, to manufacture fresh self-compacting concrete (SCC) and investigate its properties in the fresh and hardened state. Four SCC mixes were prepared by replacing natural aggregate (NA) with RCA at 0%, 25%, 50%, and 75% replacement levels. The water-cement (w/c) ratio was maintained constant at 0.38 for all the mixes. Slump Flow, J-ring, and V-funnel tests were performed on the SCC mixes in the fresh state, and the compressive strength of hardened concrete was determined after seven, 14, and 28 days. Water absorption and split tensile tests were also carried out for all the mixes. The findings revealed that it is possible to reach compressive strengths higher than 40 MPa at 28 days for RCA replacement level of 75% by using a superplasticizer and low w/c ratio. The decrease in compressive strength concerning the SCC-NA mix was 25% for 75% replacement level. The highest split tensile strength at 28 days was around 3.3 MPa for a 50% replacement level. The lowest water absorption was 3.2% for SCC-NA, which was gradually increased and was highest at 5.6% for 75% replacement level.

Keywords: recycling; self-compacting concrete (SCC); high-performance concrete (HPC); recycled concrete aggregate (RCA); natural aggregate (NA)

Citation: Katar, I.; Ibrahim, Y.; Abdul Malik, M.; Khahro, S.H. Mechanical Properties of Concrete with Recycled Concrete Aggregate and Fly Ash. *Recycling* **2021**, *6*, 23. https://doi.org/10.3390/recycling6020023

Academic Editors: Mizi Fan, Domenico Asprone and José Neves

Received: 3 January 2021
Accepted: 21 March 2021
Published: 1 April 2021

Publisher's Note: MDPI stays neutral with regard to jurisdictional claims in published maps and institutional affiliations.

1. Introduction

Recently, the main objectives of most environmental policies regarding waste are to prevent waste and promote reuse, recycling, and recovery to decrease the negative environmental impact. However, when undertaking this issue, it is essential to consider that a key barrier for developing recyclable products is the lack of demand and clients' readiness to pay more for these kinds of products, both in business to business and by consumers. Engineers have a significant role in environmentally friendly decision-making processes, as they should consider the lifelong influence and the impact on the environment [1]. Despite the mentioned barrier, RCA's use can significantly decrease concrete cost and produce an appropriate mix (located in the cost-efficient zone). As the use of a high volume of RCA decreases the mechanical characteristics of concrete, those mixes are still considered suitable when superplasticizers are used [2], and this is the case of the produced concrete with a compressive strength of around 40 MPa, which is the range for mixes used in this research.

A further dimension of the environmental requirements and customer satisfaction deals with the green quality function deployment-II (GQFD-II), a recent methodology to improve and develop products, and was successfully used by many companies. GQFD-II integrated life cycle assessment (LCA) and life cycle costing (LCC) into QFD matrices and organized customer, environmental, and cost requirements through the complete product development process. This method integrates QFD (quality), LCA (environment), and LCC (cost) into one effective tool, which could be used to evaluate the diverse product concepts considering the quality, environment, and cost [3].

The previous recycling concepts, LCA and LCC, are followed in this research to be applied to the recycled concrete aggregate (RCA) and compare it with natural aggregate (NA). The deposition of construction and demolition waste (CDW) in landfills and the conforming costs are increasing. Consequently, concrete recycling and reuse as aggregates is very advantageous and allows the volume reduction of CDW, thus reducing environmental impacts formed by the construction sector [4]. Additionally, and based on the International Organization for Standardization (ISO) definition of LCA phases, a study of life cycle inventory (LCI) assessment in Serbia concluded specific transport distances and types of impacts on total environmental impacts. The effects investigated were global warming, energy use, acidification, photochemical oxidant creation, and eutrophication [5]. The study compared two scenarios: for the first one (RCA transport distances are smaller than those of NA), the RCA and NA production environmental impact in terms of studied categories was approximately the same, and the benefit from recycling in terms of waste and natural mineral resources depletion minimizing was clearly gained. For the second one (RCA transport distances are equal to those of NA), RCA's total impacts were more significant, increasing, and ranging from 11.3% to 36.6%, depending on the impact category. Due to the study assumptions, the RCA case's energy savings were possible only for specific ratios of NA to RCA transport distances. The results accordingly stressed the necessity of getting the RCA from very close distances than NA to achieve the targeted environmental gain. The existing crusher plants are located between 100 to 200 km from Riyadh city, so the crusher for RCA situated in the city vicinity can make such concretes cost-efficient and environmentally friendly.

One of the utmost substantial advances in concrete technology is self-compacting concrete (SCC), classified as high-performance concrete. This type of concrete is produced by including add-on materials like slag cement, viscosity modifying agents, and fly ash. The SCC is requested to possess three main elementary characteristics: high resistance to segregation, restrained flowability, and high deformability [6,7]. The RCA derived from concrete waste has recently been used to supplement (NA) to manufacture fresh concrete. The shortage of NA and the rising landfill charges have taken into account RCA usage in concrete. Given RCA's underlying consistency, some researchers have shown that it may be used in the design, primarily for lower-level or non-structural applications, as an alternative to NA [8]. For the current concrete use, RCA was found to be theoretically acceptable. For barriers, shoulders, pavements, roads, embankments, bridge frames, and curbs, properly treated RCA may be included in the new concrete. It may also be found in bituminous concrete, gutters, surface bases for soil-cement, and structural grade concrete. However, the RCA obtained from demolished concrete should be seriously evaluated to pass the suitability requirements set out in the associated specifications for a particular application [9,10].

Many researchers investigated RCA beams' flexural and shear properties, RCA columns' compressive strength, and beam-column joints' seismic performance [11–13]. The effect of concrete permeability on RCA's durability in the marine environment and such aggressive environments was investigated. In these conditions, the RCA's reliability was adversely impacted by concrete permeability in the event of its decrease. By utilizing a lower water-cement (w/c) ratio, the effect of RCA's aggressive atmosphere may be minimized [14]. To guarantee fresh recycled concrete's workability, it was suggested to retain a steady amount of efficient water over the initial moisture of precise control aggregates [15].

Using RCA replacement with higher percentages in concrete (reaching 100%) and its impact on various concrete properties was considered. Concrete mixes reaching 100% RCA replacement displayed up to 37% reduction in compressive strength than natural aggregate concrete (NAC) [16]. With quality-assured conditions, concrete with a high replacement percentage of RCA from real concrete waste and laboratory waste can achieve similar splitting tensile strength, compressive strength, abrasion resistance, modulus of elasticity, and water absorption as NAC [17,18]. Considering the bond between deformed bars and concrete, it was found that concrete with NA and concrete with different replacement levels of RCA have similar bond behavior [19]. Concrete exposure to high temperature (around 600 °C) resulted in a gradual increase in specimen cracks and increased weight loss. However, these problems can be overcome by increasing the fly ash content, which reduces the loss in compressive strength and enhances RCA's ductility [20]. A similar reduction in RAC strength was experienced under intermittent and sustainable loadings [21]. The parent concrete has a considerable effect on the quality of RCA and its mechanical properties. RCA obtained from low-strength, and lightweight concrete has significantly lower ultimate bond strength and compressive strength than the one gained from a high-strength parent concrete [22].

SCC properties were studied using different replacement levels of 10%, 20%, 30%, and 40% of RCA. Flexural strength and SCC compressive strength decreased with the rise in RCA's substitution ratio [23]. RCA's mechanical properties can be improved by the appropriate use of nanomaterials, like Nanosilica and Nano-CaCO3. In some cases, higher compressive strength was obtained [24]. Combining slag, silica fume, and fly ash can gain more compressive strength to compensate for its reduction due to RCA in SCC. Using these materials in SCC with high replacement levels of RCA helps maintain resistance to segregation of this type of concrete and high filling and passing ability while reaching a comparable compressive strength that allows SCC's structural use [25]. Relating to the fine aggregate in the RCA mixture, it was observed that compressive strength of 20 MPa could be obtained from SCC when using the manufactured sand and RCA as a 100% replacement of NA, which is almost 50% of original strength when using NA. The replacement of NA with manufactured sand and RCA resulted in a reduction in split tensile strength and SCC's flexural strength [26]. Fly ash can be used with metakaolin and expanded glass aggregate for other applications that require good mechanical properties as well as thermal conductivities [27]. The use of a superplasticizer in SCC helps lower the water-cement ratio, which results in higher compressive strengths.

Based on this literature review, it is observed that RCA is already a topic of research interest, and this study aims to develop SCC mixes using RCA obtained from locally demolished structures. Managing demolition waste in Riyadh has become a critical problem in current years and is likely to escalate in the future. RCA's use at different replacement levels with a constant fly ash content is considered in this study. Therefore, this study attempts to utilize local construction demolition waste produced in Riyadh to produce high-strength concrete that can be used to construct new structures. In this research, SCC with 100% NA and three replacement levels (25%, 50%, and 75%) of RCA were produced. The fresh and hardened properties of these mixes were studied. J-ring, v-funnel, and slump flow tests were performed, and a compressive strength test after seven, 14, and 28 days were conducted. Additionally, water absorption and split tensile tests were performed on samples of all the mixes.

2. Methods

This study was conducted in different phases. In phase one, the normal SCC mix was prepared using Ordinary Portland Cement (OPC), fine aggregate, NA, fly ash, and superplasticizer. Several trials were conducted to meet the performance requirements of SCC in the fresh state. Slump flow, V-funnel, and J-ring tests were conducted to analyze the fresh properties of SCC. In the subsequent phases, the NA was replaced by 25% RCA, followed by a 50% and 75% replacement.

2.1. Preparation of Raw Materials

NA and RCA of sizes 15, 10, and 5 mm were used. RCA was obtained from the demolished concrete waste from a construction site in Riyadh. The demolished concrete was collected from the project and dumped in a different location. It was crushed manually using a steel hammer to obtain smaller size particles. It took more than a week to crush the concrete and extract the RCA's required amount and size. The RCA obtained was purely from concrete waste (concrete slab, columns, and beams), and it was ensured that it was free from other construction waste like blocks, bricks, and tiles. In the last phase, the crushed old concrete was sieved to segregate the required sizes. A sample is shown in Figure 1. RCA is inherently weaker than NA due to the presence of the old mortar layer on its surface. RCA has higher absorption because the mortar layer is porous and soft. The LA abrasion value of RCA is also higher than NA, while the specific gravity is lower. RCA's quality can be improved by minimizing the amount of attached old mortar on the surface of RCA [28].

Figure 1. Recycled coarse aggregate (RCA).

In Riyadh, OPC, fine aggregate, NA, and Type C Indian fly ash were obtained from local construction material suppliers. Class C fly ash has high cementing abilities, and these are formed from the burning of sub-bituminous coal. This type of fly ash does not need an activator (based on ASTM C 618 standards [29]) to form cementitious compounds. The specific properties of the Type C fly ash can be obtained from Mahakavi and Chithra [24].

Superplasticizer was used to get the desired properties of flowability and cohesion in the fresh state. It is a third-generation superplasticizer used to produce high-performance concrete. It allows the use of low water/binder ratios along with a high degree of compaction. It is an aqueous solution of modified polycarboxylates and co-polymers and is light brown. It may be used for all Portland cement forms, such as ground granulated blast slag, pulverized fly ash and micro-silica, and pozzolanic materials. The suggested dose is inside the binder's weight range of 0.8–2.2%. Exact dosages, though, can only be calculated by conducting trial mixes to fulfill the precise criteria. It needs to be applied to the gauging water or pumped into the concrete mixer simultaneously. The dry mix cannot be applied to it. To give the required effects, the concrete should be mixed for at least 60 s with viscocrete [30]. For the preparation of all the concrete mixes used in this study, potable tap water was used.

2.2. Concrete Mix Proportions

The concrete proportions were calculated carefully following the European Federation of National Associations Representing for Concrete (EFNARC) for SCC [31]. Additionally, a detailed literature review [19,23,24] was also conducted to finalize the mixes' proportions. Table 1 shows the SCC mix proportions.

Table 1. SCC Mix Proportions.

Ingredient (kg/m^3)	SCC-NA	SCC-RCA25%	SCC-RCA50%	SCC-RCA75%
Cement	425	425	425	425
Fly Ash	75	75	75	75
Water	191.50	191.50	191.50	191.50
w/c ratio	0.38	0.38	0.38	0.38
15 mm	300 NA	225 NA + 75 RCA	150 NA + 150 RCA	75 NCA + 225 RCA
10 mm	350 NA	262.50 NA + 87.50 RCA	175 NA + 175 RCA	87.50 NCA + 262.50 RCA
5 mm	250 NA	187.50 NA + 62.50 RCA	125 NA + 125 RCA	62.50 NCA + 187.50 RCA
Sand	840	840	840	840
Superplasticizer (L/m^3)	4 L	5 L	1.75 L	2 L

The batching of the concrete mix materials was done carefully using automated controlled weighing machines in the lab, and the weighted materials were stored in different separate buckets. Water absorption tests were conducted for NA and RCA. The water absorption was found to be 5.5%, 1%, and 0.65% for 5, 10, and 15 mm size NA, respectively, and 3.4%, 3%, and 5.4% for 5, 10, and 15 mm size RCA, respectively. Based on the water absorption results, additional water was added to the concrete mix design.

For preparing concrete mixes, a tilting drum concrete mixer has been used. The mixer's bottom was first lubricated with water to avoid sticking the material in the drum. Raw materials were then added in small quantities, along with water, while the mixer was rotated at a tilted position. Superplasticizer was dosed in small amounts to reach the required consistency. The mixer was stopped intermittently to check the concrete mix, and if found satisfactory, mixing was stopped to conduct tests on the mixes. During the initial trials, it was observed that even a slight overdose of the superplasticizer resulted in bleeding and segregation in the SCC mix. Unstable mixes were discarded, and retrials were done until all the SCC performance requirements were met as per EFNARC [31].

2.3. Testing of Fresh Properties

The slump flow and T50 measurements were used to estimate all concrete mixtures' deformability intensity and flowability. The slump flow research was carried out, as per the American Society for Measuring Materials (ASTM) C 1611 [32]. The passing capability of concrete through rebar was tested by measuring the slump-flow and T50 values using the J-Ring test following ASTM C 1621 [32]. The concrete mix's ability to resist segregation was determined by visually analyzing the concrete mix during the slump flow test and assigning a Visual Stability Index (VSI) value to each concrete mixture. Figure 2 displays the various studies performed on the most current SCC.

In the first stage, the flowability test was performed using the slump cone and steel base plate with a circle marked at the distance of 500 mm. The flowability time was recorded, and as per EFNARC standards, the SCC should touch the marked line in (2–5 s), and the slump flow diameter should range from 650 to 800 mm. During trial experiments, several attempts were made to satisfy the acceptance criteria of SCC. During the study's final experiments, the concrete total spread area and timing were recorded (Table 2). After the flowability test, a V-funnel test was performed on SCC, and as per EFNARC standards, the concrete must flow from the funnel in 6 to 12 s. The complete readings of the V-Funnel test are shown in (Table 2). In the last stage of the fresh test on SCC, the J-ring test was performed. As per the EFNARC standard, the concrete should pass through the rings at around 0 to 10 s. The J-ring tests' complete readings are shown in (Table 2).

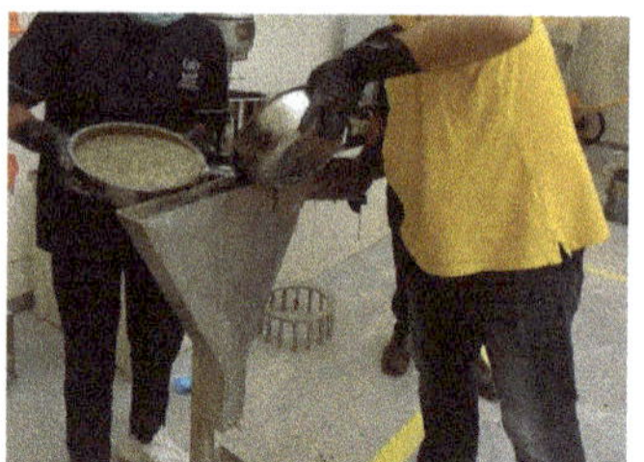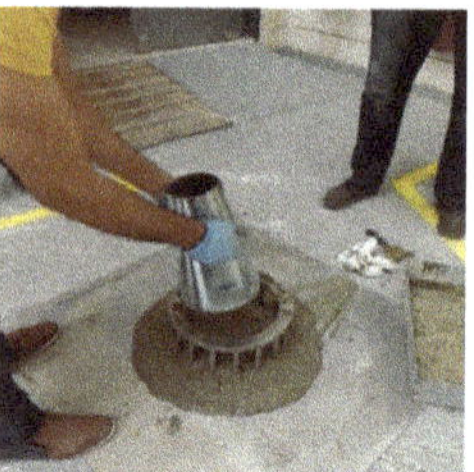

Figure 2. Flowability, V-funnel, and J-ring Tests.

Table 2. Test Results.

Mix	Slump Flow (cm)	T-50 (s)	V-Funnel (s)	J-ring (cm)	Visual Stability Index (VSI)
SCC-NCA	72	2.64	11.80	64	0 (highly stable)
SCC-RCA25%	78	2.74	8.38	72	1 (stable)
SCC-RCA50%	75	3.50	8.69	68	1 (stable)
SCC-RCA75%	66	2	6.58	56	0 (stable)

After these three tests, the SCC was cast in 12-inch standard cylinders, and twelve cylinders were cast for each trial mix in this study. The cylinders were demolded the next day, and casting tags were marked on each cylinder. Tag information includes the date of the casting, mix number, and the days for curing. In the last phase, all cylinders were placed in a curing pond at room temperature in the lab, as shown in Figure 3.

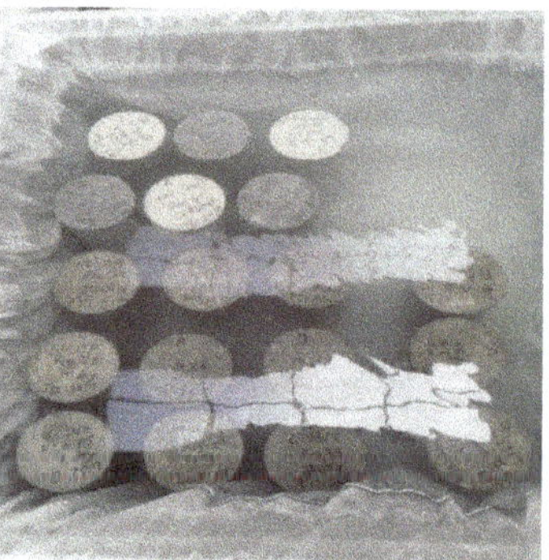

Figure 3. Concrete cylinder demolding and curing.

Potable water was used for the concrete cylinders curing in the laboratory. The curing water temperature was room temperature in the laboratory between 25–30 °C.

2.4. Testing of Hardened Properties

Nine 150 × 300 mm concrete cylinders were prepared and cured from each concrete mixture in the curing room at room temperature and at relative humidity greater than 95% before the day of testing for compressive strength testing of concrete mixtures on days 7, 14, and 28. The cylinders were inspected following ASTM C39 [32]. Two 150 × 300 mm cylinders prepared from each concrete mixture were used to perform the 28-day split tensile strength test. The splitting tensile strength of all concrete mixtures was performed by applying a diametric compressive force to a cylindrical concrete specimen placed with its horizontal axis between the plates of a measurement system in compliance with ASTM

C 496-96 [32], (C496-96). One 150×300 mm cylinder prepared from each concrete mixture was used to test water absorption.

3. Results

Table 2 shows the complete results of a various test conducted on fresh concrete, which includes slump flow, T-50 test, V-funnel, and J-ring, and it also consists of the visual stability index results:

All the mixes satisfied the criteria to qualify as SCC. In the SCC-NCA case, the superplasticizer was dosed at 0.8% of cement and fly ash weight, whereas SCC-RCA25% was dosed at 1.0%. A sudden drop in the superplasticizer dosage (0.35%) was observed in the case of SCC-RCA50%. The curing pond temperature was room temperature, and ordinary drinkable water was used for curing, as shown in Figure 3. In the subsequent phases, the cylinders were tested in a comprehensive mechanical tester to observe concrete's crushing strength, as shown in Figure 4.

Figure 4. Concrete compression test.

The concrete split tensile strength was tested using the comprehensive concrete tester with split tensile test assemblies, as shown in Figure 5.

Figure 5. Split tensile strength test.

The complete results of compressive strength and tensile strength tests and the water absorption test are shown in Table 3.

Table 3. Compressive strength and water absorption results.

Mix	Compressive Strength (MPa)			Split Tensile Strength (28 Days) (MPa)	Water Absorption (%)
	7 Day	14 Day	28 Day		
SCC-NCA	45.20	50.70	55.90	3.10	3.27
SCC-RCA25%	33.90	40.50	44.30	1.90	4.20
SCC-RCA50%	33.90	34.10	42.40	3.30	5.50
SCC-RCA75%	28.30	31.80	41.80	3	5.62

4. Discussions

Workability tests on fresh concrete mixes compressive strength at seven, 14, and 28 days and split tensile strength results are discussed in this section. The slump flow, T-50 time, V-funnel, J-ring, and VSI results are summarized in Table 2, and Figure 6 shows the slump flow for all the mixes used in this study.

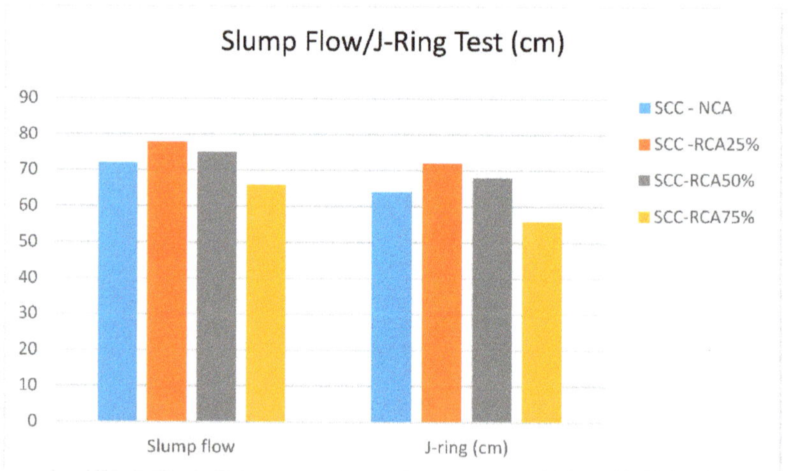

Figure 6. Slump Flow and J-Ring Test.

The slump flow was found to be within the range of 660–780 mm, within the acceptable range for SCC. The difference between slump flow diameter and J-ring test was found to be 8, 6, 7, and 10 cm, respectively, which is also within the acceptable range of 0–100 mm. The higher slump value for SCC-RCA 25% can be correlated to a higher dosage of superplasticizer. Therefore, it is essential to use the superplasticizer very carefully, and the dosage should be increased in tiny amounts to avoid high flowability. Figure 7 shows the T50 time and V-funnel time.

All the mixes satisfied the requirement of T-50 time (2–5 s) and V-funnel passing time (6–12 s), indicating good flowing and filling ability. Results from the J-ring test indicate good passing ability. The SCC mix must fulfill the fresh state criteria because it is necessary to maintain the right flowability, filling capability, and passing ability. If the SCC mix is too viscous, it can create a problem in pumping, or if it is less viscous, it might lead to segregation of the coarse aggregate and/or bleeding. This can be controlled by carefully administering the superplasticizer to the mix. Figure 8 compares the seven-, 14- and 28-days compressive strength for all the mixes designed for this study.

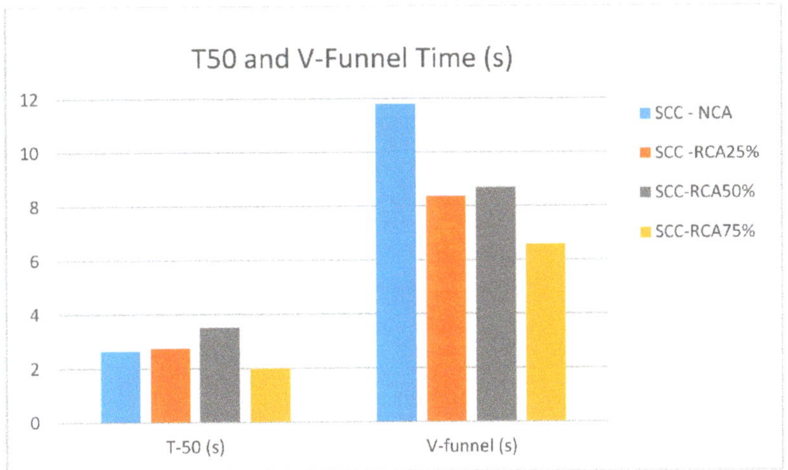

Figure 7. T50 and V-funnel Test.

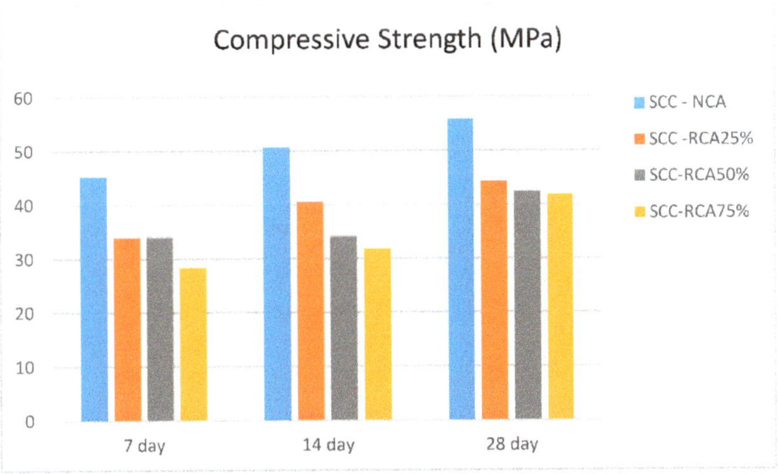

Figure 8. Compressive strength at different curing cycles.

The highest 28-day compressive strength was reported for the control mix SCC-NCA, which was 55.9 MPa. SCC-NCA and SCC-RCA25% exhibited normal strength gain, while SCC-RCA50% and SCC-RCA75% did not show considerable strength gain from seven to 14 days but had a jump in strength from 14 to 28 days. There was a reduction of 21% in the 28-day compressive strength compared to SCC-NCA at the replacement ratio of 25%, a 24.1% reduction at the replacement ratio of 50%, and a 25% reduction at the replacement level of 75%. Similar studies [16,33] have reported a reduction of 37% and 43% in compressive strength at 28 days for 100% replacement level, which is relatively higher than the results achieved in this study. Other studies [34,35] have reported a decrease of 25% at a 100% replacement level, which is lower than the present study. Hence, these research results are pretty satisfactory compared with the previous research studies of similar nature and scenarios.

SCC-RCA75% achieved almost the same 28-day strength as SCC-RCA50%. It was observed that there was a decrease in the compressive strength with an increase in RCA content.

The possible error assessment is also done using the complete samples and mixes' standard deviation test, as shown in Table 4.

Table 4. Compressive strength deviation assessment.

Mix	Compressive Strength		
	7 Day	14 Day	28 Day
SCC-NCA	34.00	51.60	59.00
	44.00	39.00	52.74
	46.00	49.80	43.74
Standard Deviation	5.25	5.56	6.26
SCC-RCA25%	34.10	43.50	44.30
	24.50	33.90	42.50
	33.80	38.20	46.10
Standard Deviation	4.46	3.93	1.47
SCC-RCA50%	38.70	32.00	40.60
	34.00	33.60	39.60
	34.00	36.10	42.40
Standard Deviation	2.22	1.69	1.16
SCC-RCA75%	26.50	33.60	39.20
	27.00	33.80	42.50
	31.70	28.00	43.90
Standard Deviation	2.34	2.69	1.97

It is observed that the maximum standard deviation reported was for SCC-NCA because of higher values of compressive strength (>35 MPa). Overall, the standard deviation values are in acceptable limits as per other studies [36]. Since the strength of the designed concrete in this study is greater than 40 MPa, it can satisfactorily be used as structural concrete [37].

A Split Tensile Strength test was conducted after 28 days of curing. Results from the split tensile strength are shown in Figure 9.

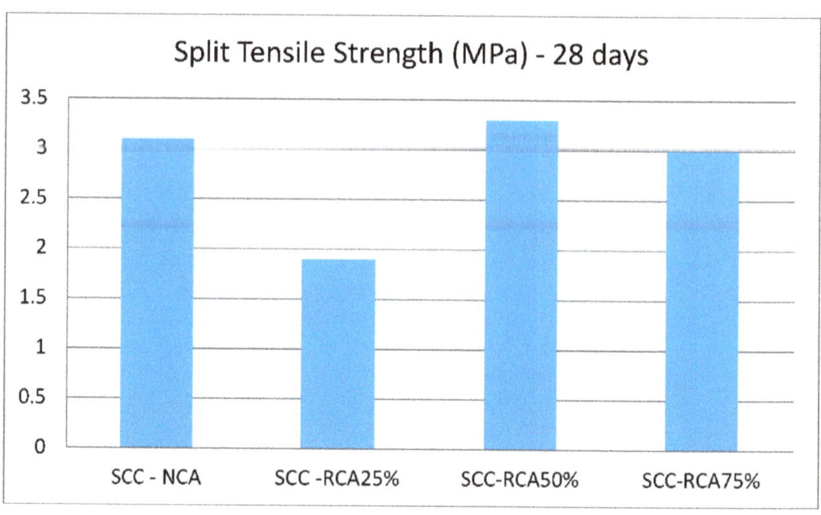

Figure 9. Split Tensile Strength of Concrete.

The values for split tensile strength were found to be in the range of 1.9–3.3 MPa, with SCC-RCA25% being the lowest at 1.9 MPa. This decrease in split tensile strength can be attributed to higher superplasticizer dosage for SCC-RCA25%, which resulted in a highly flowable mix. SCC-NCA, SCC-RCA50%, and SCC-RCA75% were 3.1, 3.3, and 3 MPa, respectively. The split tensile strength results are slightly on the lower side than concrete mixes with similar compressive strength.

5. Conclusions

The development of high-strength SCC mixes using RCA is genuinely challenging but needed. In this study, RCA obtained from the demolishing of old structures in Riyadh was used with replacement levels of 0%, 25%, 50%, and 75% forming four different SCC mixes. Fly ash and a superplasticizer were used to get the desired properties of flowability and cohesion in all mixes' fresh states. The water/cement ratio was kept constant at 0.38. The samples of each mix were tested to check the properties of fresh concrete using the slump flow test, J-ring, and V-funnel. After appropriate curing in a water tank at room temperature, cylinders of different mixes were tested to get the compressive strength, split tensile strength, and water absorption after seven, 14, and 28 days. According to the study results, the following conclusions are obtained:

1. Using RCA results in a reduction of the 7-, 14-, and 28-day compressive strength: As the replacement ratio is increased, more reduction in compressive strength is observed. The reduction of 28-day compressive strength was 21%, 24%, and 25%, for 25%, 50%, and 75% replacement levels, respectively. The minimum 28-day strength obtained was 41.8 MPa, for 75% RCA replacement, which is considered acceptable for structural applications.
2. Water absorption of the SCC is increased with the increase of the replacement level of RCA. The absorption ratio was increased by 28%, 68%, and 72%, for 25%, 50%, and 75% replacement levels, respectively.
3. There is no clear trend in the effect of the RCA replacement ratio on the split tensile strength.

RCA can enhance the environment, preserve the natural resources, and produce concrete with a reasonable compressive strength that can make the concrete eligible for use in structural applications. However, the cost reduction that resulted in using the RCA depends on crushing concrete plants within a reasonable distance to the site, which may encourage and promote more RCA usage that can improve the environment.

Author Contributions: Conceptualization: I.K., and Y.I.; methodology: M.A.M., and S.H.K.; formal analysis: M.A.M. and S.H.K.; investigation: I.K., and Y.I.; writing—original draft preparation: Y.I.; writing—review and editing: I.K., M.A.M., and S.H.K.; supervision: Y.I.; project administration: Y.I.; funding acquisition: Y.I. All authors have read and agreed to the published version of the manuscript.

Funding: This research was funded by Prince Sultan University, Seed Project of the 2020/2021 academic year.

Institutional Review Board Statement: Not applicable.

Informed Consent Statement: Not applicable.

Data Availability Statement: Not applicable.

Acknowledgments: The authors would like to thank PSU for funding this project. The authors would also like to thank Engineer Omar Shabir for his efforts to conduct the tests and obtain the results. Finally, the authors thank the Structures and Materials Research Lab at the College of Engineering for its support.

Conflicts of Interest: The authors declare no conflict of interest.

References

1. Pajunen, N.; Rintala, L.; Aromaa, J.; Heiskanen, K. Recycling—The importance of understanding the complexity of the issue. *Int. J. Sustain. Eng.* **2016**, *9*, 93–106. [CrossRef]
2. Rawaz, K.; Jorge, B.; José, D. Combined Economic and Mechanical Performance Optimization of Recycled Aggregate Concrete with High Volume of Fly Ash. *Appl. Sci.* **2018**, *8*, 1189. [CrossRef]
3. Zhang, Y. Green QFD-II: A life cycle approach for environmentally conscious manufacturing by integrating LCA and LCC into QFD matrices. *Int. J. Prod. Res.* **1999**, *37*, 1075–1091. [CrossRef]
4. Estanqueiro, B.; Silvestre, J.D.; de Brito, J.; Pinheiro, M.D. Environmental life cycle assessment of coarse natural and recycled aggregates for concrete. *Eur. J. Environ. Civ. Eng.* **2018**, *22*, 429–449. [CrossRef]
5. Marinković, S.; Radonjanin, V.; Malešev, M.; Ignjatovic, I. Comparative environmental assessment of natural and recycled aggregate concrete. *Waste Manag.* **2010**, *30*, 2255–2264. [CrossRef]
6. Khatib, J.M. Performance of self-compacting concrete containing fly ash. *Constr. Build. Mater.* **2008**, *229*, 1963–1971. [CrossRef]
7. Hossain, K.M.A.; Lachemi, M. Fresh, mechanical, and durability characteristics of self-consolidating concrete incorporating volcanic ash. *J. Mater. Civ. Eng.* **2010**, *227*, 651–657. [CrossRef]
8. Hwang, S.D.; Khayat, K.H.; Bonneau, O. Performance-Based Specifications of Self-Consolidating Concrete Used in Structural Applications. *ACI Mater. J.* **2006**, *103*, 121.
9. Grdic, Z.J.; Toplicic, C.G.A.; Despotovic, I.M.; Ristic, N.S. Properties of self-compacting concrete prepared with coarse recycled concrete aggregate. *Constr. Build. Mater.* **2010**, *24*, 1129–1133. [CrossRef]
10. Kou, S.; Poon, C. Properties of self-compacting concrete prepared with coarse and fine recycled concrete aggregates. *Cem. Concr. Compos.* **2009**, *31*, 622–627. [CrossRef]
11. DG/TJ07-008. *Technical Code of Application of Recycled Aggregate Concrete*; Shanghai Construction Standard Society (SCSS): Shanghai, China, 2007.
12. Thomas, C.; Setién, J.; Polanco, J.A.; Cimentada, A.I.; Medina, C. Influence of curing conditions on recycled aggregates concrete. *Constr. Build. Mater.* **2018**, *72*, 618–625. [CrossRef]
13. Montero, J.; Laserna, S. Influence of effective mixing water in recycled concrete. *Constr. Build. Mater.* **2017**, *132*, 343–352. [CrossRef]
14. Alexandridou, C.; Angelopoulos, G.N.; Coutelieris, F.A. Mechanical and durability performance of concrete produced with recycled aggregates from Greek construction and demolition waste plants. *J. Clean. Prod.* **2018**, *176*, 745–757. [CrossRef]
15. Pedro, D.; de Brito, J.; Evangelista, L. Structural concrete with simultaneous incorporation of fine and coarse recycled concrete aggregates: Mechanical, durability and long-term properties. *Constr. Build. Mater.* **2017**, *154*, 294–309. [CrossRef]
16. Bui, N.K.; Satomi, T.; Takahashi, H. Improvement of mechanical properties of recycled aggregate concrete basing on a new combination method between recycled aggregate and natural aggregate. *Constr. Build. Mater.* **2017**, *148*, 376–385. [CrossRef]
17. Shi, Z.L. Experimental study on recycled aggregate concrete and its engineering application. *China Fly Ash* **2004**, *4*, 3–4. (In Chinese)
18. Panda, K.C.; Bal, K.C. Properties of self-compacting concrete using recycled coarse aggregate. In Proceedings of the Chemical, Civil and Mechanical Engineering Tracks of 3rd Nirma University, International Conference on Engineering (NUiCONE), Ahmedabad, India, 6–8 December 2012.
19. Omrane, M.; Kenai, S.; Kadri, E.-H.; Ait-Mokhtar, A.K. Performance and durability of self-compacting concrete using recycled concrete aggregate and natural pozzolan. *J. Clean. Prod.* **2017**, *165*, 415–430. [CrossRef]
20. Govind, G.; Bhupinder, S. Analytical investigation in bond of deformed steel bars in recycled aggregate concrete. *J. Sustain. Cem. Based Mater.* **2020**. [CrossRef]
21. Qianqian, R.; Yaopeng, W.; Xu, Z.; Yonghui, W. Effects of fly ash on the mechanical and impact properties of recycled aggregate concrete after exposure to high temperature. *Eur. J. Environ. Civ. Eng.* **2019**. [CrossRef]
22. Zhiyu, L.; Wengui, L.; Vivian, W.Y.T.; Jianzhuang, X.; Surendra, P.S. Current progress on nanotechnology application in recycled aggregate concrete. *J. Sustain. Cem. Based Mater.* **2019**, *8*, 79–96. [CrossRef]
23. Zhanggen, G.; Jing, Z.; Tao, J.; Tianxun, J.; Chen, C.; Rui, B.; Yan, S. Development of sustainable self-compacting concrete using recycled concrete aggregate and fly ash, slag, silica fume. *Eur. J. Environ. Civ. Eng.* **2020**. [CrossRef]
24. Mahakavi, P.; Chithra, R. Effect of recycled coarse aggregate and manufactured sand in self-compacting concrete. *Aust. J. Struct. Eng.* **2020**, *21*, 33–43. [CrossRef]
25. Bin, L.; Wengui, L.; Zhiyu, L.; Xitao, L.; Vivian, W.Y.T.; Zhuo, T. Performance deterioration of sustainable recycled aggregate concrete under combined cyclic loading and environmental actions. *J. Sustain. Cem. Based Mater.* **2020**. [CrossRef]
26. Assaad, J.J.; Matar, P.; Gergess, A. Effect of quality of recycled aggregates on bond strength between concrete and embedded steel reinforcement. *J. Sustain. Cem. Based Mater.* **2020**, *9*, 94–111. [CrossRef]
27. Longo, F.; Cascardi, A.; Lassandro, P.; Aiello, M.A. A new Fabric Reinforced Geopolymer Mortar (FRGM) with mechanical and energy benefits. *Fibers* **2020**, *8*, 49. [CrossRef]
28. Verian, K.P.; Ashraf, W.; Cao, Y. Properties of recycled concrete aggregate and their influence in new concrete production. Resources. *Conserv. Recycl.* **2018**, *133*, 30–49. [CrossRef]
29. ASTM. *Standard Specification for Coal Fly Ash and Raw or calcined Natural Pozzolan for Use in Concrete*; ASTM Standard C618; ASTM: West Conshohocken, PA, USA, 2012.

30. Product Data Sheet. Sika Viscocrete—021301011000001467. 2020. Available online: https://gcc.sika.com/content/dam/dms/gcc/o/sika_viscocrete_ts-100.pdf (accessed on 22 March 2021).
31. European Federation of National Associations Representing for Concrete (EFNARC). *Specifications and Guidelines for Self-Compacting Concrete*; EFNARC: Surrey, UK, 2002.
32. ASTM C1611/C1611M-18. American Society for Testing and Materials, Standards and Publications. Available online: https://www.astm.org/Standard/standards-and-publications.html (accessed on 10 July 2018).
33. Mohammed, D.; Tobeia, S.; Mohammed, F.; Hasan, S. *Compressive Strength Improvement for Recycled Concrete Aggregate*; Building and Construction Engineering Department, University of Technology: Baghdad, Iraq, 2018.
34. Khan, A.R.; Fareed, S.; Khan, M.S. Use of Recycled Concrete Aggregates in Structural Concrete. In Proceedings of the Fifth International Conference on Sustainable Construction Materials and Technologies, London, UK, 14–17 July 2019.
35. Qasrawi, H.; Marie, I. Towards Better Understanding of Concrete Containing Recycled Concrete Aggregate. *Adv. Mater. Sci. Eng.* **2013**, *2013*, 636034. [CrossRef]
36. Zhang, X.B.; Fang, Z.; Deng, S.C. Study on the standard deviation for the compressive strength of recycled concrete. *Adv. Mater. Res.* **2013**, *639–640*, 313–318. [CrossRef]
37. ACI 318. Building Code Requirements for Structural Concrete. Available online: https://engineervincentpardopilien.weebly.com/uploads/2/1/5/1/21511442/aci_318-2011.pdf (accessed on 22 March 2021).

Article

The Effect of Recycled HDPE Plastic Additions on Concrete Performance

Tamrin [1,*] and Juli Nurdiana [1,2]

1 Faculty of Engineering, Mulawarman University, Samarinda 75117, Indonesia; j.nurdiana@utwente.nl
2 Department of Governance and Technology for Sustainability, University of Twente, 7522 NB Enschede, The Netherlands
* Correspondence: fts_tamrin@ft.unmul.ac.id

Abstract: This study examined HDPE (high-density polyethylene) plastic waste as an added material for concrete mixtures. The selection of HDPE was based on its increased strength, hardness, and resistance to high temperatures compared with other plastics. It focused on how HDPE plastic can be used as an additive in concrete to increase its tensile strength and compressive strength. 156 specimens were used to identify the effect of adding different percentages and sizes of HDPE lamellar particles to lower, medium, and higher strength concrete for non-structural applications. HDPE 0.5 mm thick lamellar particles with sizes of 10×10 mm, 5×20 mm, and 2.5×40 mm were added at 2.5%, 5%, 10%, and 20% by weight of cement. The results showed that the medium concrete class (with compressive strength equal to 10 MPa) had the best response to the addition of HDPE. The 5% HDPE addition represented the optimal mix for all concrete types, while the 5×20 mm size was best.

Keywords: concrete additive; concrete mixture; plastic waste; HDPE; plastic lamellar particles

Citation: Tamrin; Nurdiana, J. The Effect of Recycled HDPE Plastic Additions on Concrete Performance. *Recycling* **2021**, *6*, 18. https://doi.org/10.3390/recycling6010018

Received: 17 October 2020
Accepted: 3 March 2021
Published: 6 March 2021

Publisher's Note: MDPI stays neutral with regard to jurisdictional claims in published maps and institutional affiliations.

1. Introduction

Plastic has long been considered a manmade material with many benefits. It has lightweight properties and is easily shaped to the designer's desires. Its versatile properties have led to its widespread use. Since 2016–2017, plastic consumption has increased from 335 million tons to 348 million tons. This demand is expected to reach 485 million tons by 2030 [1]. The downside of plastic use is the waste generated and the environmental pollution caused because many plastics are not biodegradable and can take between 500 and 1000 years to decompose [2]. The pollution risks from the toxins released can impact groundwater quality, animal/human health, food-chain poisoning, and reduction in soil fertility [3]. Furthermore, if burnt in an open space, plastics produce carbon monoxide (a greenhouse gas). If disposed of in waterways, plastics can cause siltation and impede water flows, thereby creating a flood risk [4,5]. Research on beaches has shown that coastline plastic waste in 192 countries in 2010 amounted to between 4.8 and 12.7 million metric tons [6]. This waste threatens marine organisms [7] and has led to many demands to restrict plastic use and reshape behavior at the consumer level [8]. Recycling has increased in developed countries since 2006 [1] and offers a partial solution. The regular process of plastic recycling starts with sorting it into several polymer types, followed by cleaning, scraping, smelting, and converting it into pellets to be repurposed into plastic bags, plastic containers, carpets, jacket insulation, and other materials. However, traditional recycling suffers from cross-contamination and requires high energy consumption [9]. In 2018, processing plastic waste for energy used 43% of all of the collected post-consumer waste stream [1]. Furthermore, the insufficient processing and management of plastic waste worldwide face the challenge of insufficient plastic waste treatment facilities at all stages of collection, separation, and disposal. By 2050, it is projected that about 12 billion metric tons of plastic litter will end up in landfills and the natural environment [10]. Many countries, including Indonesia, experience problems with plastic waste. Indonesia generates some 67.8 million tons of waste, with plastic waste being the second-largest waste stream after organic waste, reaching 17% in 2018 [11]. To solve the problems of plastic waste and divert this away from landfill, requires any opportunities to be identified within the value chains.

Investing in a circular system to manage plastic pollution offers potential solutions with social and environmental benefits. Circularity will retain the value of plastic materials if they are returned back into the supply chain, thus reducing the volume of discarded plastics ending up in nature. Therefore, the identification of a relevant local strategy for waste (including plastics) and the tailoring of partnerships to suit various stakeholders (i.e., businesses, industries, and civil society) are necessary [12,13]. Here, building a nexus between the waste and construction sectors emerges as a possible option for increasing plastic circularity, especially macro-plastics, which are in widespread use [14]. The additional value to be obtained from their use as an additive in concrete mixtures could also create new business opportunities [15]. The final application from plastic additions to concrete, as examined in this study is expected to be for non-structural projects, such as wall panels, parking lots, or paths [16–18]. Even plastic fibers can be used below the concrete layer in constructing rigid pavements.

Concrete has properties that are sensitive to the type of added materials that are beyond those specified in the traditional job mix design. The strength of concrete depends on the type and size of the aggregates used [19–21], and different additive materials produce variations in tensile strength and compressive strength [22–25]. Single-use plastics are considered suitable for disposal as admixtures in concrete, as low-carbon reusable materials, e.g., PET (polyethylene terephthalate) [26] and HDPE (high-density polyethylene) [27]. The advantages of using plastic additions in concrete are that they are lightweight, better resistant to weather, waterproof [28], and confer thermal insulation properties [9,29]. However, compared to PET, HDPE has higher temperature resistance than PET (melting at 130–135 °C). Further, as Merli et al. [30] identified, HDPE is less discussed in the literature compared to PET. This motivated our interest and focus on HDPE.

A few researchers have discussed the use of HDPE in concrete in different contexts. For instance, Pesic et al. [31] investigated the effect recycled HDPE fibers had for reinforced concrete for structural uses using two different fiber diameters (0.25 mm and 0.4 mm) with 0.40%, 0.75%, and 1.25% fiber volume fraction. The study showed that the HDPE fiber reinforced concrete of 0.75–1.25% could maintain a constant post-cracking tensile of 30–40% of the flexural peak capacity. The use of HDPE for non-structural uses was discussed by Lopez et al. [32], who considered using recycled HDPE as a partial replacement of coarse aggregate in mixes of Acrylic Polymer Pervious Concrete (AcPPC) at ratios of 10%, 20%, and 30% at sizes of $\frac{1}{2}''$ and $\frac{3}{4}''$. The study showed that the optimum strength was reached by a 10% addition at sizes of $\frac{1}{2}''$. Further, by using a different type of plastic, Jain et al. [16] investigated the effect of plastic bag additions to concrete at 0.5, 1, 2, 3, and 5% of the weight of concrete. They found that a higher percentage of plastic reduced concrete's workability. The addition also affected the bonding between plastic aggregate and cement paste, as it created voids, thus reducing the concrete strength. The above-mentioned studies clearly emphasized that plastic addition could benefit concrete properties at certain levels, and contribute to sustainable construction. However, how different types of plastics may affect the behavior of concrete is an interesting issue, which offers scope for discussion and development.

Unlike these previous studies, we have investigated different aspects. Our study examines the potential use of HDPE addition on different concrete classes. We assess the effect of various HDPE lamellar particle sizes and percentages as lightweight admixtures into different concrete mixes used for non-structural works, but not as a replacement to cement or other materials. This paper is structured as follows: Section 1 is an introduction providing the background and aim of this study. It is followed by a description of materials and methods in Section 2. The results of the tests are provided in Section 3 and discussed in Section 4. Conclusions and recommendations for future research are presented in Section 5.

2. Materials and Methods

This study used concrete mixes formed from cement and aggregate (fine and coarse aggregate). These are designed to fall into three concrete classes: lower, medium, and higher concrete strength. Lower concrete strength is in the following named as B0 and it represents concrete with cylindrical strength of f′c = 7 MPa. Medium and high concrete strength here refers to cylindrical strength f′c equal to 10 MPa and 25 MPa respectively. Three different sizes of HDPE lamellar (10 × 10 mm, 5 × 20 mm, and 2.5 × 40 mm) with the same thickness of 0.5 mm were added to the mixtures to examine their effect on concrete properties. The ACI (American Concrete Institute) and ASTM (American Society for Testing and Materials) testing standards were used to calculate specific gravity, slump value, unit weight, tensile and compressive strength. Table 1 provides a summary of the standard testing used in this research.

Table 1. The standards used for concrete testing.

Standard	Targeted Testing
ASTM C-127	Specific gravity of coarse aggregate
ASTM C33-99a	Adequate grading requirement and aggregate quality; sieve analysis
ASTM C29/C29M-07	Unit weight for fine and coarse aggregate
ASTM C131/C131M-20	Resistance to degradation by abrasion on small-size coarse aggregate
ACI 211.1-91	Standard Practice for Selecting Proportions for Normal, Heavyweight, and Mass Concrete
ASTM C143	Slump test
ASTM C39	Compressive strength
ASTM C496	Tensile strength

2.1. Materials

2.1.1. Cement

As this study's scope involved non-structural applications, the examination used cement type 1, which is intended for walls, pavement, sidewalks, and other precast products. Using the ASTM C-127 standard, this cement material was found to have a specific gravity of 3.18 g/cm^3, which falls in the acceptable range of 3.1–3.3 g/cm^3. This cement composition comprises four main chemical compounds, i.e., tricalcium silicate ($3CaO \cdot SiO_2$), shortened to C_3S (55 wt.%), dicalcium silicate ($2CaO \cdot SiO_2$), abbreviated to C_2S (17 wt.%), tricalcium aluminate ($3CaO \cdot Al_2O_3$), shortened to C_3A (10 wt.%), tetracalcium aluminoferrite ($4CaO \cdot Al_2O_3 \cdot Fe_2O_3$), shortened to C_4AF (7 wt.%), carbon disulphide (CS_2) (6 wt.%). In addition, there are small amounts of minor compounds, e.g., alkali (Na_2O), free calcium oxide (free CaO), ignition loss, and magnesium oxide (MgO) of which, according to Indonesian national standard (SNI No 15-2049/2015), the maximal amount should be less than 5 and 6 wt.% respectively.

2.1.2. The Aggregates

The aggregates refer to any particulates used as an inert filler in concrete. These vary from sand, gravel, crushed stone to blast-furnace slag. Following ASTM C33, the aggregates are categorized into fine and coarse aggregate. This study used sand as a fine aggregate within a range of 0.1–10 mm (Figure 1a), and crushed stone as a coarse aggregate meeting the range of 2–30 mm (Figure 1b). These aggregates were collected from Palu, Central Sulawesi, Indonesia. Palu's aggregates are considered basalt and are widely used for lightweight building walls and concrete in Indonesia. Its physical characteristics and quality provided adequate consolidation in concrete mixes, and offers higher resistance to alkali-silica reaction, compared to other aggregates obtained from areas in East Kalimantan.

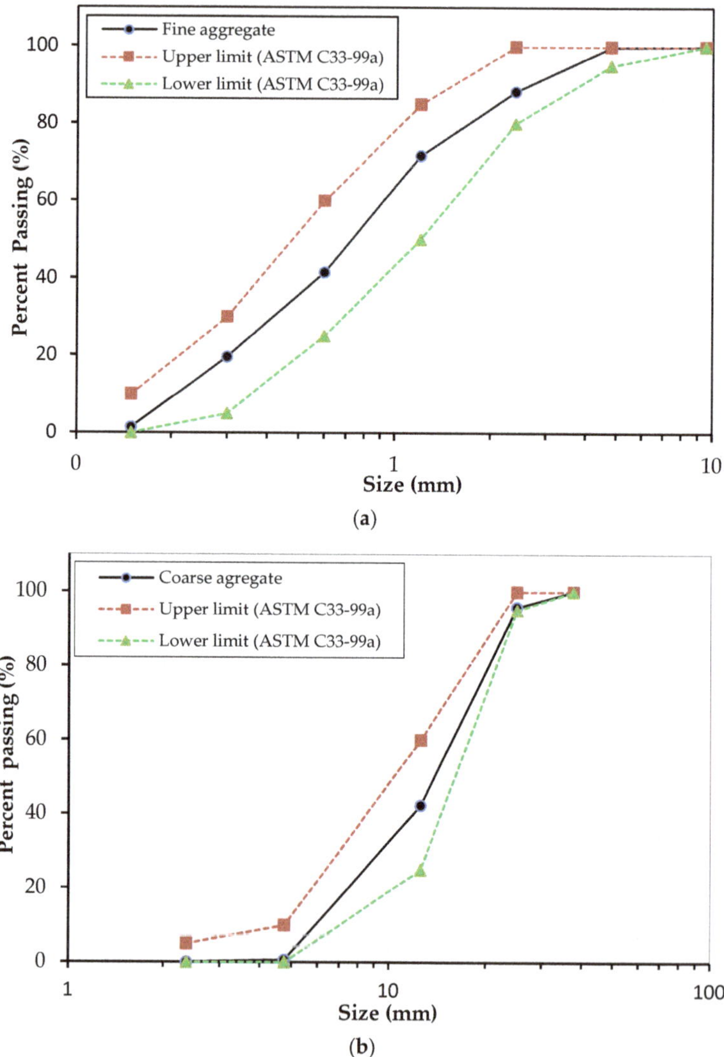

Figure 1. Aggregate size: (**a**) fine aggregate; (**b**) coarse aggregate.

These aggregates were tested at the Faculty of Engineering Laboratory, Mulawarman University, Samarinda, following the ASTM C33-99a standard [33] for sieve analysis, which defines the adequate grading requirement and aggregate quality in concrete. The results of this test for fine and coarse aggregate are shown in Figure 1.

The unit weight testing of fine and coarse aggregates was conducted using the ASTM C29/C29M-07 standard [34]. 2.55 g/cm^3 for the coarse aggregate and 2.54 g/cm^3 for the fine aggregate was obtained, meeting the standard requirement of 2.5–2.7 g/cm^3. The test for coarse aggregate abrasion was conducted using the ASTM C131/C131M-20 standard [35] and a result of 23% was obtained. This value was lower than the 27% ASTM limit. Table 2 provides the detailed physical properties of the materials used for mixing concrete.

Table 2. Physical properties of cement and aggregates.

Materials	Properties	Value	Allowable Range
Cement type 1	Specific gravity	3.18 g/cm^3	3.1–3.3 g/cm^3
Fine Aggregate	Sieve size	Figure 1a	Following ASTM C33-99a
	Unit weight	2.54 g/cm^3	2.5–2.7 g/cm^3
Coarse Aggregate	Sieve size	Figure 1b	Following ASTM C33-99a
	Unit weight	2.55 g/cm^3	2.5–2.7 g/cm^3
	Resistance to abrasion	23%	Maximum of 27%

2.1.3. Specimen Preparation of HDPE Lamellar Particles

The HDPE plastic materials were collected from wastes disposed of in Samarinda landfills to reflect potential future plans to reduce non-sustainable waste that contaminate waterways and aquifers. They were rinsed in preparation for the cutting process. Figure 2a shows the production of lamellar particles and how we ensured a similar thickness for all the sheets. We implemented a cutting procedure using markings determined as a function of the pattern and size. First, a selection of HDPE plastic samples 0.5 mm thick was cut into lamellar particles before adding to the concrete mixture, excluding any thicker or less than 0.5 mm. This process produced three sizes, namely, 10 × 10 mm, 5 × 20 mm, and 2.5 × 40 mm, each with an identical surface area of 1 cm^2. This ensures commonality of interaction between the plastic addition and cement in any change in propertie00s of concrete mixture and the bonding effect. Figure 2b shows an example of HDPE lamellar particles with a size of 10 × 10 mm at a similar thickness of 0.5 mm after the cutting process.

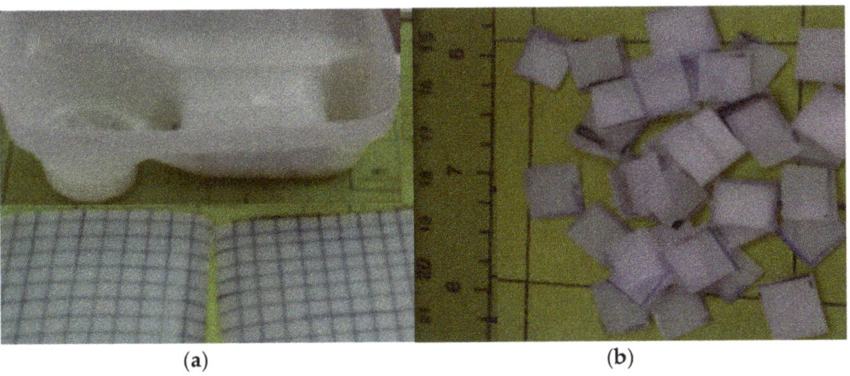

(a)   (b)

Figure 2. The preparation of high-density polyethylene (HDPE) addition: (**a**) marking procedure for the cutting process; (**b**) examples of HDPE lamellar particles with a size of 10 × 10 mm.

2.2. Concrete Preparation and Testing

2.2.1. Job Mix Design

The concrete mix design and the material composition of the three concrete types are shown in Table 3. The process of identifying the right proportion of concrete mixture complied with the standard practice for selecting proportions for normal, heavyweight, and mass concrete (ACI 211.1-91) [36]. Therefore, concrete strength tests were performed 28 days after casting to ensure that the resultant properties satisfied quality control designs. The concrete specimens were demolded after 24 h and kept in a water curing tank until the age of testing at a room temperature of 27 °C. We defined the slump value of the mixture for the three different concrete classes at the same range so it met cost efficiency and workability in the field. To meet this value, we set the w/c ratio at different levels, according to the water to cement ratio used.

Table 3. Concrete job mix design.

Description	B0	f′c10	f′c25
Targeted average compressive strength of the concrete	7 MPa	10 MPa	25 MPa
Water to cement ratio	0.95	0.63	0.52
Slump value	120 ± 5 mm	120 ± 5 mm	120 ± 5 mm
Amount of water	180 kg/m^3	190 kg/m^3	215 kg/m^3
Amount of cement	190 kg/m^3	295 kg/m^3	413 kg/m^3
Fine aggregate content	969 kg/m^3	828 kg/m^3	687 kg/m^3
Coarse aggregate content	1010 kg/m^3	1014 kg/m^3	1220 kg/m^3

2.2.2. Mixing Process

Table 4 shows the design experiments for the three concrete types, four different percentages of HDPE additions, lamellar particle sizes, and various aggregate particles used for the mixtures as described in Figure 1. The process started by mixing the different cement types and aggregates under dry conditions for a few minutes before adding water. The lamellar particles were then added to each concrete type according to their size categories (10×10 mm; 5×20 mm; 2.5×40 mm) until the concrete mixture became homogeneous.

Table 4. Design experiment of specimens used.

HDPE Addition	Volume of Concrete (m^3)	Cement (kg)	Fine Aggregate (kg)	Coarse Aggregate (kg)	Water (kg)	HDPE Lamellar (kg)	Number of Specimens Compressive Strength	Number of Specimens Tensile Strength
B0	0.021	4.03	20.54	21.41	3.82	0	2	2
B0-HDPE 2.5%								
10×10 mm	0.021	4.03	20.54	21.41	3.82	0.10	2	2
5×20 mm	0.021	4.03	20.54	21.41	3.82	0.10	2	2
2.5×40 mm	0.021	4.03	20.54	21.41	3.82	0.10	2	2
B0-HDPE 5%								
10×10 mm	0.021	4.03	20.54	21.41	3.82	0.20	2	2
5×20 mm	0.021	4.03	20.54	21.41	3.82	0.20	2	2
2.5×40 mm	0.021	4.03	20.54	21.41	3.82	0.20	2	2
B0-HDPE 10%								
10×10 mm	0.021	4.03	20.54	21.41	3.82	0.40	2	2
5×20 mm	0.021	4.03	20.54	21.41	3.82	0.40	2	2
2.5×40 mm	0.021	4.03	20.54	21.41	3.82	0.40	2	2
B0-HDPE 20%								
10×10 mm	0.021	4.03	20.54	21.41	3.82	0.81	2	2
5×20 mm	0.021	4.03	20.54	21.41	3.82	0.81	2	2
2.5×40 mm	0.021	4.03	20.54	21.41	3.82	0.81	2	2
f′c10	0.021	6.25	17.55	21.49	4.03	0	2	2
f′c10-HDPE 2.5%								
10×10 mm	0.021	6.25	17.55	21.49	4.03	0.16	2	2
5×20 mm	0.021	6.25	17.55	21.49	4.03	0.16	2	2
2.5×40 mm	0.021	6.25	17.55	21.49	4.03	0.16	2	2
f′c10-HDPE 5%								
10×10 mm	0.021	6.25	17.55	21.49	4.03	0.31	2	2
5×20 mm	0.021	6.25	17.55	21.49	4.03	0.31	2	2
2.5×40 mm	0.021	6.25	17.55	21.49	4.03	0.31	2	2

Table 4. *Cont.*

HDPE Addition	Volume of Concrete (m³)	Cement (kg)	Fine Aggregate (kg)	Coarse Aggregate (kg)	Water (kg)	HDPE Lamellar (kg)	Number of Specimens	
							Compressive Strength	Tensile Strength
f'c10-HDPE 10%								
10 × 10 mm	0.021	6.25	17.55	21.49	4.03	0.63	2	2
5 × 20 mm	0.021	6.25	17.55	21.49	4.03	0.63	2	2
2.5 × 40 mm	0.021	6.25	17.55	21.49	4.03	0.63	2	2
f'c10-HDPE 20%								
10 × 10 mm	0.021	6.25	17.55	21.49	4.03	1.25	2	2
5 × 20 mm	0.021	6.25	17.55	21.49	4.03	1.25	2	2
2.5 × 40 mm	0.021	6.25	17.55	21.49	4.03	1.25	2	2
f'c25	0.021	8.75	14.56	25.09	4.56	0.00	2	2
f'c25-HDPE 2.5%								
10 × 10 mm	0.021	8.75	14.56	25.09	4.56	0.22	2	2
5 × 20 mm	0.021	8.75	14.56	25.09	4.56	0.22	2	2
2.5 × 40 mm	0.021	8.75	14.56	25.09	4.56	0.22	2	2
f'c25-HDPE 5%								
10 × 10 mm	0.021	8.75	14.56	25.09	4.56	0.44	2	2
5 × 20 mm	0.021	8.75	14.56	25.09	4.56	0.44	2	2
2.5 × 40 mm	0.021	8.75	14.56	25.09	4.56	0.44	2	2
f'c25-HDPE 10%								
10 × 10 mm	0.021	8.75	14.56	25.09	4.56	0.88	2	2
5 × 20 mm	0.021	8.75	14.56	25.09	4.56	0.88	2	2
2.5 × 40 mm	0.021	8.75	14.56	25.09	4.56	0.88	2	2
f'c25-HDPE 20%								
10 × 10 mm	0.021	8.75	14.56	25.09	4.56	1.75	2	2
5 × 20 mm	0.021	8.75	14.56	25.09	4.56	1.75	2	2
2.5 × 40 mm	0.021	8.75	14.56	25.09	4.56	1.75	2	2

The terms used are as follows: B0 refers to normal concrete meeting the standard job mix design without the addition of HDPE lamellar, while B0-HDPE 2.5% refers to B0 concrete with the addition of 2.5% HDPE. The amount of HDPE lamellar particles for the experimental investigation is calculated on the basis of the weight of the cement used.

This study used cylindrical specimens with a diameter of 150 mm and a height of 300 mm (Figure 3). The cylinder molds were made of steel to avoid leakage and hold their integrity under severe use. The mold nonabsorbent material avoids a reaction with Portland or other hydraulic cement. For each test, two samples were used for each size of HDPE lamellar particles. Accordingly, the number of samples used for the splitting tensile and cylindrical compressive strength tests was six. The total number of samples used was 156, including those for normal concrete testing. Since only two specimens were used for each design, the data were processed as average.

We set a higher water/cement ratio to produce a workable concrete (minimum 0.52). Typically, the minimum water/cement ratio is 0.35–0.4, as a lower ratio may result in the concrete becoming too dry and unworkable [37]. Furthermore, the use of a higher water/cement ratio results in a high slump value. However, the addition of HDPE plastic sheets compensated for this change. To evaluate the effect of adding HDPE lamellar particles, several tests were conducted, including slump testing using the ASTM C143 standard [38], compressive strength testing using the ASTM C39 standard [39], and tensile strength testing using the ASTM C496 standard [40].

Figure 3. Preparation of concrete cylinder specimen.

3. The Results

This study conducted tests to examine the appropriate concrete mixtures incorporating HDPE lamellar particles for non-structural applications to determine the effect of HDPE size and additions on low-quality concrete, medium-quality concrete, and high-quality concrete can be explained as follows.

3.1. Concrete Slump Test

Concrete workability is quantified by the concrete slump, which depends on many factors, e.g., mixing methods, concrete materials and admixtures, and the workability changes with time due to those factors. In this slump test, the preparation of specimens using the mold (slump cone) is shown in Figure 4, and the varying HDPE lamellar particles were added to the fresh concrete before testing.

Figure 4. Preparation of B0 concrete for the slump test.

We set the slump value for normal concrete (baseline) to 115–125 mm. As shown in Figure 5, a greater amount of HDPE to the concrete mix led to a smaller slump value. The slump value of B0 concrete (Figure 5a) with HDPE size at 10 × 10 mm declined by 10% for 20% HDPE lamellar addition (accounted from 120 mm of normal concrete to 110 mm). The lowest percentage reduction at slump value showed up in f′c10 with HDPE size 10 × 10 mm (Figure 5b), and the maximum value given by f′c25 with HDPE size 2.5 × 40 mm, at 16.7% (Figure 5c). We found, from all the samples used, the maximum value of reduction ranged from 5 to 20 mm compared to the standard value.

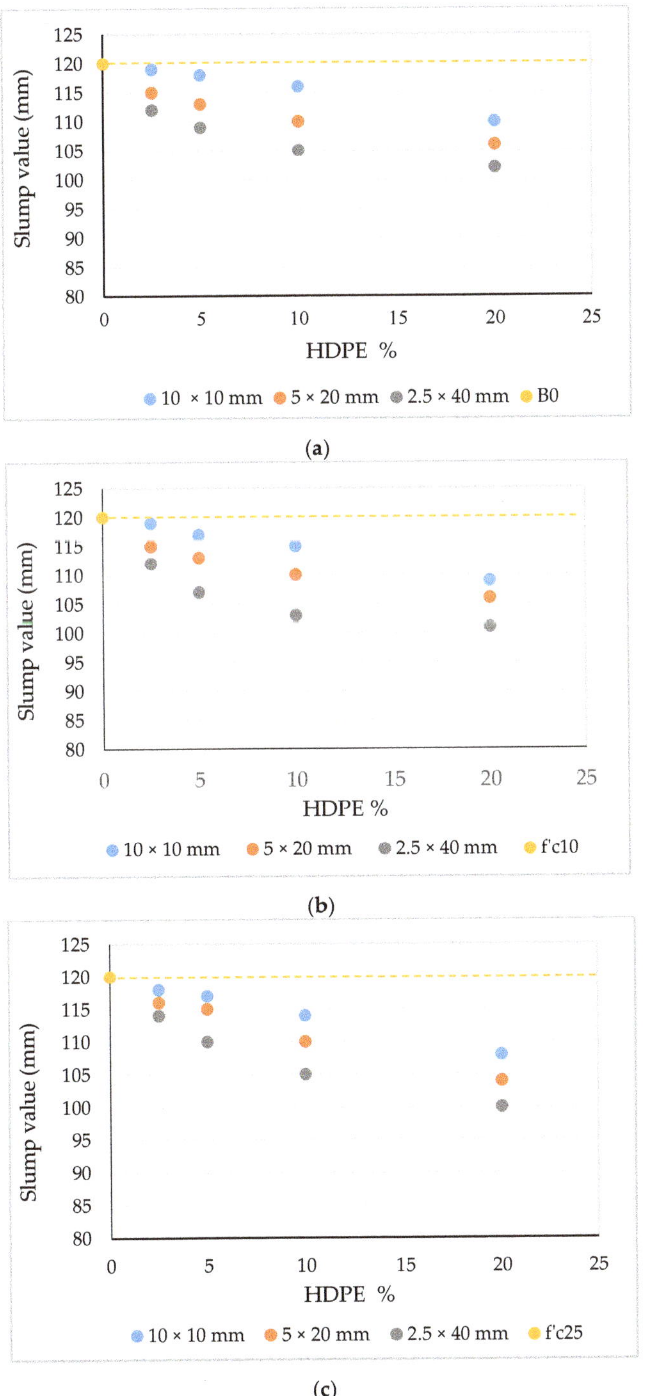

Figure 5. Slump value as a function of HDPE content (%) and sizes: (**a**) B0; (**b**) f′c10; (**c**) f′c25.

3.2. Unit Weight of Concrete

ASTM C29 [34] uses the term unit weight to refer to the concrete property in mass per unit volume. This gives a good indication of sample concrete density. The unit weight for all samples was determined by comparing the specimen's weight with the specimen's substantial volume. The relationship between the unit weight of the concrete, HDPE lamellar content, and its sizes are given in Figure 6. The graphs show that a greater addition of HDPE lamellar particles led to lighter concrete due to the low density of HDPE plastic, which was applied to B0, f′c10, and f′c25. However, the size of the HDPE sheets did not affect the concrete unit weight, as they all showed a similar value for certain percentages. For example, in Figure 6b, the unit weight of 20% additions was 2011 kg/m^3, 2013 kg/m^3, and 2012 kg/m^3 for HDPE sizes of "10 × 10 mm", "5 × 20 mm", and "2.5 × 40 mm" respectively.

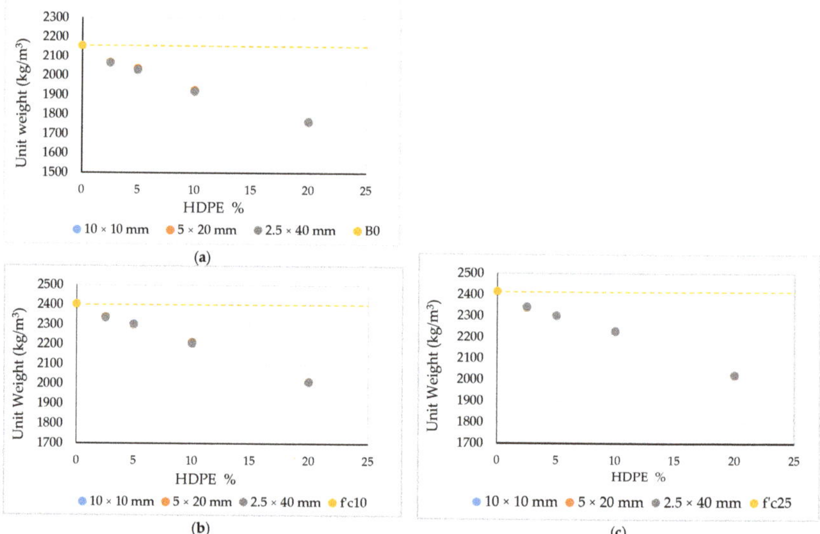

Figure 6. The relationship between concrete unit weight and HDPE content and sizes: (**a**) B0; (**b**) f′c10; (**c**) f′c25.

3.3. Tensile and Compressive Strength

Tensile and compressive strengths are the important mechanical properties that identify concrete performance. Figures 7 and 8 display the results of splitting tensile and cylindrical compressive strength tests for the concrete mixtures containing HDPE addition (the testing results are shown in Appendix A). Our experiments indicated that the strength varied depending on the HDPE content and sizes.

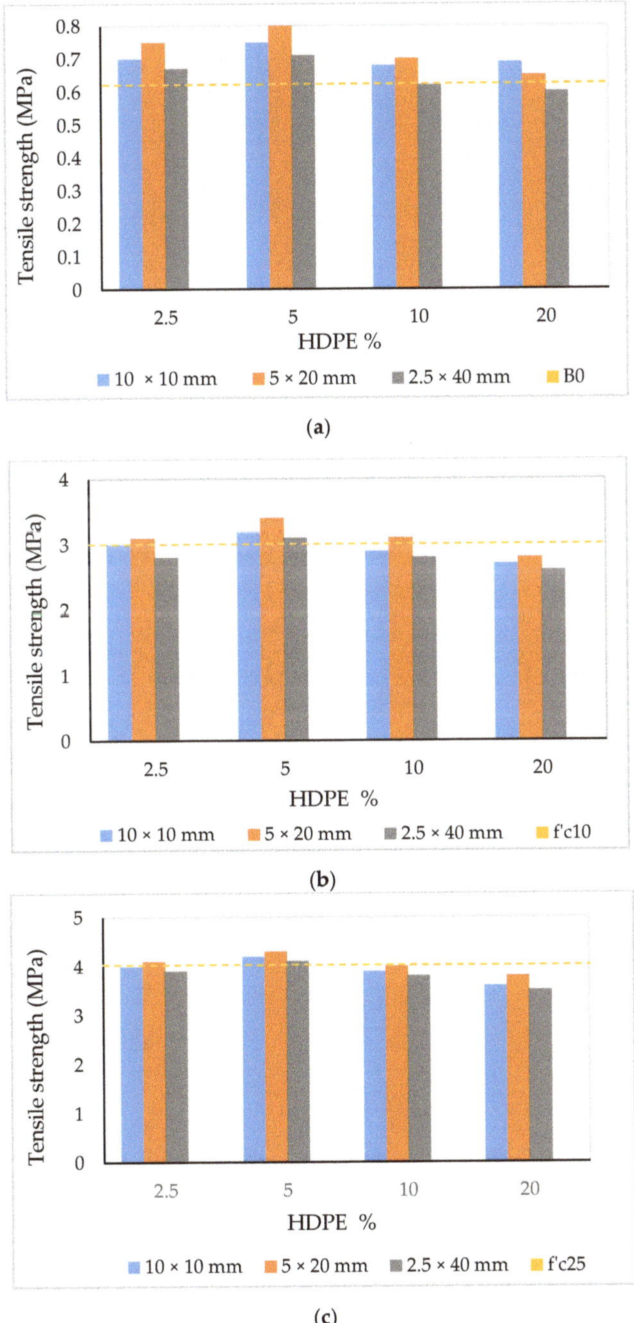

Figure 7. Tensile strength of concrete mixtures as a function of HDPE content and lamellar shape: (**a**) B0; (**b**) f′c10 MPa; (**c**) f′c25 MPa.

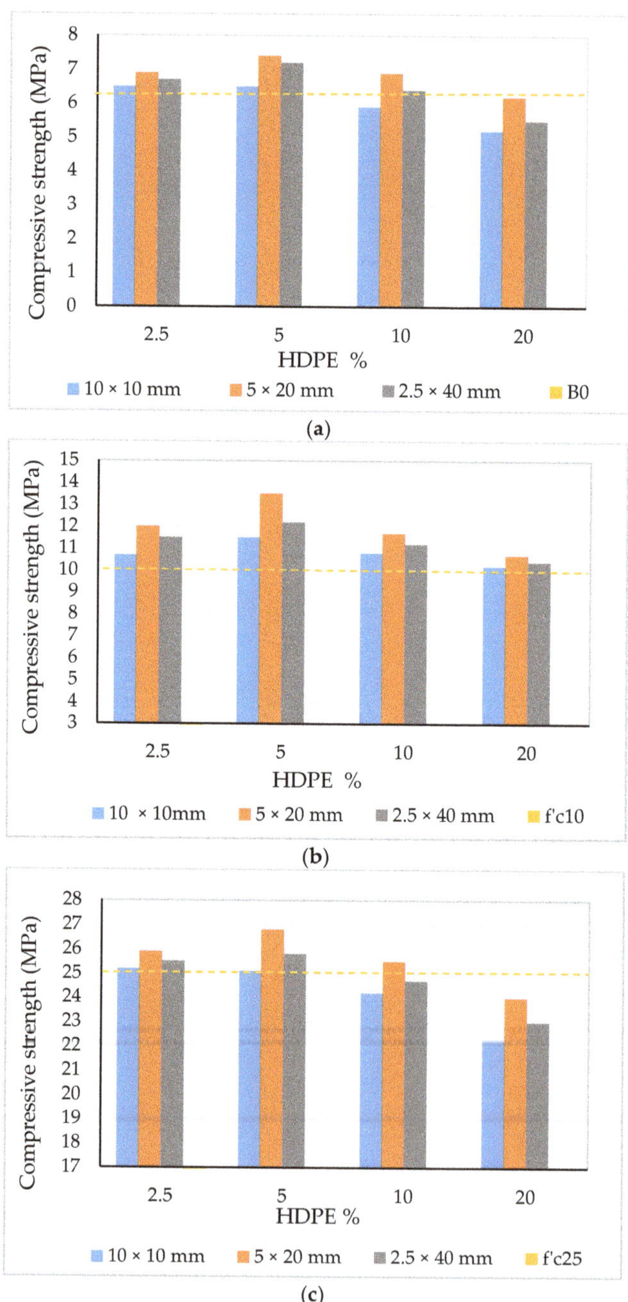

Figure 8. Compressive strength of concrete mixtures as a function of HDPE content and lamellar shape: (**a**) B0; (**b**) f'c10 MPa; (**c**) f'c25 MPa.

In line with the job mix design shown in Table 3, the concrete aggregate content and w/c ratio had an effect on tensile and compressive strength. Figure 7 shows that the B0 concrete had a lower tensile strength, compared to f'c10 and f'c25. We observed also that the percentage and HDPE sizes behaved differently on concrete strength. Although all sizes of HDPE lamellar had an equal surface area (1 cm^2), they produced a different response. In this case, we found that for all concrete classes, the addition of HDPE up to 5% showed higher tensile strength compared to other percentages (above the baseline), and the "5 × 20 mm" was the best shape compared to the size of "10 × 10 mm" and "2.5 × 40 mm". This finding aligned with the compressive strength results shown in Figure 8a–c, where the 5% addition and "5 × 20 mm" strength value was above the normal concrete quality. This graph also indicates that f′c10 concrete gave a better response to the increase in concrete quality compared to B0 and f′c25, where in all percentage additions and different lamellar shapes, the value was above the normal concrete.

In addition, for the compaction, Figure 9a,b shows a two-dimensional (2D) image of HDPE positions in the concrete mixture. The lamellar particles with sizes of "10 × 10 mm" and "5 × 20 mm" remained unaffected by coarse aggregate pressure during casting. When tested for compressive strength, the broken piece of concrete was then observed visually. The size of 10 × 10 mm and 5 × 20 mm seemed to pack and bond together with the concrete mixture in a straight position. However, we found that the "2.5 × 40 mm" sheet became curved during casting (Figure 9c). Though this condition depends on the different angles of lamellar particles and the level of coarse aggregate pressure received during casting, the results showed that the 5 × 20 mm specimens performed better compared to 2.5 × 40 mm.

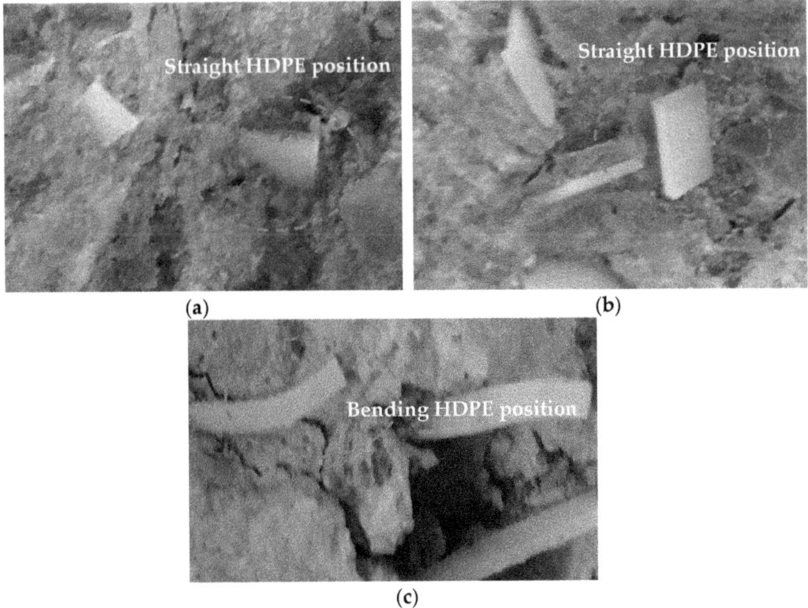

(a)　　　　　　　　　　　(b)

(c)

Figure 9. 2D images of HDPE in concrete: (**a**) 10 × 10 mm; (**b**) 5 × 20 mm; (**c**) 2.5 × 40 mm.

4. Discussion and Analysis

4.1. Relationship between HDPE Additions and Slump Value

The concrete workability indicates the consistency of the concrete mix during the work. It relates to the degree of its compaction provided by the external (contact with the surface) and internal friction (given by the aggregate size, shape, grading). The use of admixtures, e.g., plastics could also affect the compaction. The fact that plastic materials are generally lightweight and resistant to weather, means that they can be considered as suitable additive materials for concrete [19]. Previous studies indicated that added materials, including plastics, can improve the properties of concrete given appropriate percentage mixes [9,28].

In this case, the addition of plastic can improve toughness and energy absorption at post-cracking [41,42]. Further, given the very poor biodegradability of plastic, i.e., HDPE, can not only improve the long-term performance of concrete structures but it can also contribute to environmental sustainability and performance of the construction industry [43,44].

However, the addition of HDPE to the concrete mixture affected the slump value, which is essential for concrete workability. Good workability is associated with the finishing stages, homogeneity, and resistance to segregation. In regard to segregation, this must be avoided during casting, due to fresh concrete's low workability [45]. Increasing the amount of water used in the job mix and adding materials at proportioning mix to maintain concrete density [46] could be useful. For instance, recently, two types of added plastic have been used in concrete mixes and have shown satisfactory results when using 30% plastic waste in the total aggregate [47]. Therefore, to prevent less fluidity in the first place, we managed the w/c ratio and admixture additions that function to alter concrete properties when making the job mixes. Table 3 shows we set a higher w/c ratio than the typical w/c ratio, and we used a lamellar shape (bigger size than fiber).

Even so, our findings, shown in Figure 5, identified that a small amount of added plastic does not affect the mixture's workability, but a higher percentage of added plastic was found to influence the cementitious materials' content and sharply decrease the concrete's workability. The possible reason is plastic hydrophobicity, which causes an insufficiency of plastics to mix with other materials. It may limit the hydration of cement as a result of the lesser bonding between plastic surfaces and cement paste. Additionally, due to the angular and nonuniform nature of plastic aggregate particles [28,48], the increase of HDPE lamellar particles in the concrete mixture results in lower fluidity compared to normal concrete. This applies to different sizes of plastic lamellar particles, as seen in Figure 5, which clearly indicates that, for all additions of HDPE at different sizes, the addition of HDPE lamellar particles thickened the concrete mixture, thereby lowering the slump value. To sum up, when considering the range of slump reduction, the added HDPE lamellar particles fit well for low-degree workability applications. We consider that the findings of this study may be useful for non-structural works, where higher strength is not the main aspect [49,50].

4.2. Relationship between HDPE Additions and Unit Weight

When the HDPE lamellar particles were added into the concrete mixture, due to the nature of plastic (e.g., immiscibility), the addition of plastic in the mixture may increase the air content in concrete, thus affecting its density [19]. Additionally, during the hydration and curing process, some amount of moisture is removed from the concrete mixture. The immiscibility of plastics could affect compactness when a certain level of admixture is replaced by HDPE. However, here, the volume of concrete remains the same, as indicated by the amount pushed out of the sample mold. Further, as the unit weight of normal concrete is about 2400 kg/m^3; higher than HDPE, which is about 930–970 kg/m^3, it is justifiable that concrete containing HDPE additions would have lower unit weight. Figure 6 shows that for all concrete classes, compared to normal concrete, the unit weight of the concrete mixture dropped linearly with the increased number of lamellar particles in the mixture. Thus, an increase in HDPE content enables a reduction in concrete weight; an important target in construction. In particular, for the same percentage of HDPE lamellar additions, the different size of plastic lamellar particles does not affect the unit weight. This finding is indicated by an almost similar value of unit weight for "5 × 20 mm", "2.5 × 40 mm", and "10 × 10 mm" lamellar particles.

Further, this study identified that the addition of HDPE, in lamellar shape, provided the best response to concrete quality up to 5 % for medium concrete strength of f'c10. This type of concrete could add benefit to non-structural walls, base concrete in the rigid pavement on highways, paving blocks for parking lots with low loads, wall panels, shotcrete (or Gunite), and concrete footpaths. For the use of precast concrete walls especially, concrete mixtures containing HDPE could reduce the building's structural load and energy consumption within the building by lowering the inside temperature. Together with fillers (e.g., sand, quarry fine), this type of concrete mix could help prevent heat transfer within a structure, which is relevant to Indonesia's moderate to high temperatures. In particular, there is a strong connection between thermal conductivity and concrete's substantial weight whereby the use of plastics to replace aggregates can reduce concrete's thermal conductivity compared to bare concrete [9,51].

Although the previous study has discussed the development of lightweight concrete using HDPE additions (25%), opening up new development opportunities for non-structural and structural applications [52], our findings show that the concrete with added plastics should be directed to medium concrete strength. Therefore, mean concrete containing plastics cannot be used as a primary construction material, i.e., for column, beam, and plate constructions, mainly due to safety factors and its poor fire-resistant behavior [50]. These findings have an impact and add to the development of lightweight concrete for the green construction sector. Thus, the addition of HDPE could also lead to a more sustainable approach to reducing plastic waste.

4.3. The Effect of HDPE Additions to Tensile and Compressive Strength

As stated earlier, previous studies have found a relationship between plastics addition influencing concrete's tensile strength and compressive strength [19,21,53] matching those of steel fibers that affected the value of splitting tensile and compressive strength [54], as well as plastic fibers, carbon fibers, and fibers from natural materials, such as flax or other plants. According to Hasan et al. [41], inserting fibers into a concrete mixture can increase the concrete composite's tensile strength by about 10–15%, compared to standard concrete. Other research found similar results that fibers can prevent brittle failure and enhance the ductility of the concrete [48,52,54].

The tensile strength is an essential determinant of how concrete performs under induced stress. Figures 7 and 8 show the connection between tensile and compressive strength; although their relationship is not directly proportional. The higher the compressive strength, the higher the tensile strength, but at a decreasing rate [55]. This study indicates that the addition of 5% HDPE increases the tensile and compressive strength of concrete, better than 2.5%, 10%, and 20%. Furthermore, the addition of 10% and 20% HDPE content to B0 and f'c25 concrete reduced the tensile and compressive strength. This finding in line with other studies showing that increasing the volume fraction can affect fiber bonding and decrease the strength of concrete composites [15,25,51]. However, this study identified that this did not apply to f'c10 MPa concrete, where an increase in quality occurred, even with a content of 20% for sheets with a size of 5 × 20 mm (13% increase in splitting tensile strength; 35% increase in compressive strength). Therefore, the amount of added HDPE should be chosen on the basis of the weight of the cement used, as outlined in Table 2.

Further, since all HDPE samples added had the same cross-sectional area, the size largely determines the results, whereby the position of plastic lamellar particles in the concrete can reduce the optimality of the aggregate bond, as seen in Figure 9. The findings emphasized that even though the plastic lamellar particles have the same cross-sectional area, different results came because their position depends on the pressure received. This explains why some lamellar particles are curved and some are straight. Figures 7 and 8 show that in particular for lamellar particles at a size of "10 × 10 mm", the plastic lamellar particles are packed together in a straight position, but in terms of its capacity to withstand loads, the size of "5 × 20 mm" produced a better response compared to "10 × 10 mm". It shows that the length of the lamellar particles is important, up to a certain size. In summary, the performance of the additions with respect to strength testing was in the order of 5 × 20 mm > 2.5 × 40 mm > 10 × 10 mm. Thus, the use of HDPE with a size of 5 × 20 mm as an additive in the concrete mixture was acceptable.

5. Conclusions and Recommendations for Future Research

Few studies have assessed the effect of added particles' length on concrete properties. This study has contributed to the understanding of the optimal percentages and sizes of HDPE in the shape of lamellar particles in concrete. Our study contributes to showing the effect that HDPE additions have in terms of size and percentage on concrete qualities to improve its use and exploitation and to design the concrete mix design process. Some important findings are:

(1). This study evaluated the use of 2.5%, 5%, 10%, and 20% HDPE lamellar particle additions at sizes of 10 × 10 mm, 0.5 × 20 mm, and 2.5 × 40 mm incorporated into three concrete types (B0, f'c10, and f'c25). The f'c10 MPa concrete performed best in response to the addition of lamellar particles, whereas 5% was the optimal HDPE content, and 5 × 20 mm was the optimal size.

(2). All variants of HDPE lamellar particles described can be used with f'c10 MPa concrete. However, only 5 × 20 mm HDPE sheets should be used with B0 and f'c25 MPa concrete.

(3). Future research should investigate f'c10 MPa to determine the effects of different percentage additions and material composition into concrete mixes. Additionally, further work is needed to identify whether similar effects apply to different plastic shapes. More testing could explore the valuation of physical concrete properties, e.g., water porosity.

Author Contributions: T.; conceptualization, design and analysis, investigation, initial draft, visualization; J.N.; draft preparation, administration process, and the editing process. All authors have read and agreed to the published version of the manuscript.

Funding: This research received no external funding.

Data Availability Statement: All data is contained within the article.

Acknowledgments: We thank our colleagues in the Department of Civil Engineering, Mulawarman University, who help and provide us their insight and suggestions that greatly assisted our research.

Conflicts of Interest: The authors declare no conflict of interest associated with this publication. There has been no significant financial support for this work that could have influenced its outcome.

Appendix A

Table A1. Compressive and Tensile Test Results of 156 Specimens Used.

HDPE Addition	Compressive Test (MPa)			Tensile Test (MPa)		
	Number of Specimens			Number of Specimens		
	1	2	Average	1	2	Average
B0	6.40	6.30	6.35	0.60	0.68	0.64
B0-HDPE 2.5%						
10 × 10 mm	6.30	6.70	6.50	0.70	0.70	0.70
5 × 20 mm	6.80	7.00	6.90	0.80	0.70	0.75
2.5 × 40 mm	6.80	6.60	6.70	0.70	0.65	0.68
B0-HDPE 5%						
10 × 10 mm	6.40	6.60	6.50	0.70	0.80	0.75
5 × 20 mm	7.40	7.40	7.40	0.80	0.80	0.80
2.5 × 40 mm	7.15	7.25	7.20	0.70	0.74	0.72
B0-HDPE 10%						
10 × 10 mm	5.90	6.00	5.95	0.70	0.65	0.68
5 × 20 mm	6.90	6.90	6.90	0.70	0.70	0.70
2.5 × 40 mm	6.40	6.42	6.41	0.64	0.60	0.62
B0-HDPE 20%						
10 × 10 mm	5.20	5.20	5.20	0.69	0.69	0.69
5 × 20 mm	6.20	6.20	6.20	0.65	0.66	0.66
2.5 × 40 mm	5.50	5.45	5.48	0.60	0.60	0.60

Table A1. *Cont.*

HDPE Addition	Compressive Test (MPa)			Tensile Test (MPa)		
	Number of Specimens			Number of Specimens		
	1	2	Average	1	2	Average
f′c10	10.05	10.00	10.03	3.00	3.00	3.00
f′c10-HDPE 2.5%						
10 × 10 mm	10.70	10.70	10.70	2.80	3.20	3.00
5 × 20 mm	12.00	12.00	12.00	2.90	3.30	3.10
2.5 × 40 mm	11.30	11.70	11.50	2.70	2.90	2.80
f′c10-HDPE 5%						
10 × 10 mm	11.50	11.50	11.50	3.20	3.15	3.18
5 × 20 mm	13.50	13.50	13.50	3.30	3.45	3.38
2.5 × 40 mm	12.20	12.17	12.19	3.10	3.15	3.13
f′c10-HDPE 10%						
10 × 10 mm	10.80	10.80	10.80	2.80	3.00	2.90
5 × 20 mm	11.71	11.68	11.70	3.20	3.00	3.10
2.5 × 40 mm	11.20	11.20	11.20	2.65	2.90	2.78
f′c10-HDPE 20%						
10 × 10 mm	10.20	10.20	10.20	2.60	2.80	2.70
5 × 20 mm	10.68	10.72	10.70	2.80	2.85	2.83
2.5 × 40 mm	10.40	10.40	10.40	2.60	2.70	2.65
f′c25	25.00	25.10	25.05	4.00	4.10	4.05
f′c25-HDPE 2.5%						
10 × 10 mm	25.40	25.00	25.20	4.10	4.00	4.05
5 × 20 mm	25.50	26.30	25.90	4.15	4.10	4.13
2.5 × 40 mm	25.30	25.70	25.50	3.80	3.98	3.89
f′c25-HDPE 5%						
10 × 10 mm	25.00	25.20	25.10	4.20	4.15	4.18
5 × 20 mm	26.60	27.00	26.80	4.30	4.35	4.33
2.5 × 40 mm	26.10	25.50	25.80	4.20	4.10	4.15
f′c25-HDPE 10%						
10 × 10 mm	24.40	24.00	24.20	3.80	3.90	3.85
5 × 20 mm	25.60	25.40	25.50	4.00	3.90	3.95
2.5 × 40 mm	24.40	25.00	24.70	3.90	3.70	3.80
f′c25-HDPE 20%						
10 × 10 mm	22.60	22.00	22.30	3.70	3.60	3.65
5 × 20 mm	24.50	23.60	24.05	3.90	3.72	3.81
2.5 × 40 mm	22.90	23.10	23.00	3.50	3.52	3.51

References

1. The Association of Plastic Recyclers (APR): Recognition Program Operating Procedures. Available online: https://plasticsrecycling.org/images/pdf/Recognition_Program/Procedure/Recognition_ProgramOperating_Procedures_June_2009.pdf (accessed on 27 May 2020).
2. How Long It Takes for Some Everyday Items to Decompose. Available online: http://storage.neic.org/event/docs/1129/how_long_does_it_take_garbage_to_decompose.pdf (accessed on 27 May 2020).
3. Verma, R.; Vinoda, K.; Papireddy, M.; Gowda, A. Toxic Pollutants from Plastic Waste—A Review. *Procedia Environ. Sci.* **2016**, *35*, 701–708. [CrossRef]
4. Lebreton, L.C.M.; Van Der Zwet, J.; Damsteeg, J.-W.; Slat, B.; Andrady, A.; Reisser, J. River plastic emissions to the world's oceans. *Nat. Commun.* **2017**, *8*, 15611. [CrossRef] [PubMed]
5. Van Emmerik, T.; Schwarz, A. Plastic debris in rivers. *Wiley Interdiscip. Rev. Water* **2020**, *7*, e1398. [CrossRef]
6. Jambeck, J.R.; Geyer, R.; Wilcox, C.; Siegler, T.R.; Perryman, M.; Andrady, A.; Narayan, R.; Law, K.L. Plastic waste inputs from land into the ocean. *Science* **2015**, *347*, 768–771. [CrossRef] [PubMed]
7. Mattsson, K.; Hansson, L.-A.; Cedervall, T. Nano-plastics in the aquatic environment. *Environ. Sci. Process. Impacts* **2015**, *17*, 1712–1721. [CrossRef] [PubMed]
8. Baier, D.; Rausch, T.M.; Wagner, T.F. The Drivers of Sustainable Apparel and Sportswear Consumption: A Segmented Kano Perspective. *Sustainability* **2020**, *12*, 2788. [CrossRef]
9. Poonyakan, A.; Rachakornkij, M.; Wecharatana, M.; Smittakorn, W. Potential Use of Plastic Wastes for Low Thermal Conductivity Concrete. *Materials* **2018**, *11*, 1938. [CrossRef]

10. Geyer, R.; Jambeck, J.R.; Law, K.L. Production, use, and fate of all plastics ever made. *Sci. Adv.* **2017**, *3*, e1700782. [CrossRef]
11. Sistem Informasi Pengelolaan Sampah Nasional. Available online: http://sipsn.menlhk.go.id/?q=3a-komposisi-sampah (accessed on 4 August 2020).
12. Godfrey, L. Waste Plastic, the Challenge Facing Developing Countries—Ban It, Change It, Collect It? *Recycling* **2019**, *4*, 3. [CrossRef]
13. Nurdiana, J.; Franco-García, M.-L.; Hophmayer-Tokich, S. Incorporating circular sustainability principles in DKI. Jakarta: Lessons learned from Dutch business schools management. In *Towards Zero Waste*; Franco-García, M.L., Carpio-Aguilar, J., Bressers, H., Eds.; Greening of Industry Networks Studies; Springer: Cham, Switzerland, 2019; Volume 6, pp. 145–163.
14. Napper, I.E.; Thompson, R.C. Plastic Debris in the Marine Environment: History and Future Challenges. *Glob. Chall.* **2020**, *4*, 1900081. [CrossRef]
15. EMF (The Ellen Mac Arthur Foundation). Urban Biocycles. 2017. Available online: https://www.ellenmacarthurfoundation.org/publications/urban-biocyles (accessed on 20 June 2020).
16. Jain, A.; Siddique, S.; Gupta, T.; Jain, S.; Sharma, R.K.; Chaudhary, S. Fresh, Strength, Durability and Microstructural Properties of Shredded Waste Plastic Concrete. *Iran. J. Sci. Technol. Trans. Civ. Eng.* **2019**, *43*, 455–465. [CrossRef]
17. Kaufmann, J.; Frech, K.; Schuetz, P.; Münch, B. Rebound and orientation of fibers in wet sprayed concrete applications. *Constr. Build. Mater.* **2013**, *49*, 15–22. [CrossRef]
18. Alani, A.M.; Beckett, D. Mechanical properties of a large scale synthetic fibre reinforced concrete ground slab. *Constr. Build. Mater.* **2013**, *41*, 335–344. [CrossRef]
19. Babafemi, A.J.; Šavija, B.; Paul, S.C.; Anggraini, V. Engineering Properties of Concrete with Waste Recycled Plastic: A Review. *Sustainability* **2018**, *10*, 3875. [CrossRef]
20. Islam, J.; Meherier, S.; Islam, A.R. Effects of waste PET as coarse aggregate on the fresh and harden properties of concrete. *Constr. Build. Mater.* **2016**, *125*, 946–951. [CrossRef]
21. Batayneh, M.; Marie, I.; Asi, I. Use of selected waste materials in concrete mixes. *Waste Manag.* **2007**, *27*, 1870–1876. [CrossRef]
22. Akinpelu, M.A.; Odeyemi, S.O.; Olafusi, O.S.; Muhammed, F.Z. Evaluation of splitting tensile and compressive strength relationship of self-compacting concrete. *J. King Saud Univ.-Eng. Sci.* **2019**, *31*, 19–25. [CrossRef]
23. Lavanya, G.; Jegan, J. Evaluation of relationship between split tensile strength and compressive strength for geopolymer concrete of varying grades and molarity. *Int. J. Appl. Eng. Res.* **2015**, *10*, 35523–35529.
24. Choi, Y.; Yuan, R.L. Experimental relationship between splitting tensile strength and compressive strength of GFRC and PFRC. *Cem. Concr. Res.* **2005**, *35*, 1587–1591. [CrossRef]
25. Kim, S.B.; Yi, N.H.; Kim, H.Y.; Kim, J.-H.J.; Song, Y.-C. Material and structural performance evaluation of recycled PET fiber reinforced concrete. *Cem. Concr. Compos.* **2010**, *32*, 232–240. [CrossRef]
26. Nikbin, I.M.; Rahimi, S.; Allahyari, H.; Fallah, F. Feasibility study of waste Poly Ethylene Terephthalate (PET) particles as aggregate replacement for acid erosion of sustainable structural normal and lightweight concrete. *J. Clean. Prod.* **2016**, *126*, 108–117. [CrossRef]
27. Silva, A.L.P.; Prata, J.C.; Walker, T.R.; Campos, D.; Duarte, A.C.; Soares, A.M.; Barceló, D.; Rocha-Santos, T. Rethinking and optimising plastic waste management under COVID-19 pandemic: Policy solutions based on redesign and reduction of single-use plastics and personal protective equipment. *Sci. Total Environ.* **2020**, *742*, 140565. [CrossRef] [PubMed]
28. Bahij, S.; Omary, S.; Feugeas, F.; Faqiri, A. Fresh and hardened properties of concrete containing different forms of plastic waste—A review. *Waste Manag.* **2020**, *113*, 157–175. [CrossRef] [PubMed]
29. Fraternali, F.; Ciancia, V.; Chechile, R.; Rizzano, G.; Feo, L.; Incarnato, L. Experimental study of the thermo-mechanical properties of recycled PET fiber-reinforced concrete. *Compos. Struct.* **2011**, *93*, 2368–2374. [CrossRef]
30. Merli, R.; Preziosi, M.; Acampora, A.; Lucchetti, M.C.; Petrucci, E. Recycled fibers in reinforced concrete: A systematic literature review. *J. Clean. Prod.* **2020**, *248*, 119207. [CrossRef]
31. Pešić, N.; Živanović, S.; Garcia, R.; Papastergiou, P. Mechanical properties of concrete reinforced with recycled HDPE plastic fibres. *Constr. Build. Mater.* **2016**, *115*, 362–370. [CrossRef]
32. Lopez, N.; Collado, E.; Diacos, L.A.; Morente, H.D. Evaluation of Pervious Concrete Utilizing Recycled HDPE as Partial Replacement of Coarse Aggregate with Acrylic as Additive. *MATEC Web Conf.* **2019**, *258*, 01018. [CrossRef]
33. ASTM C 33-99ae1. *Standard Specification for Concrete Aggregates*; ASTM International: West Conshohocken, PA, USA, 2002.
34. ASTM International. *Standard Test Method for Bulk Density ("Unit Weight") and Voids in Aggregate*; ASTM C29/C29M-07: West Conshohocken, PA, USA, 2003.
35. ASTM International. *Standard Test Method for Resistance to Degradation of Small-Size Coarse Aggregate by Abrasion and Impact in the Los Angeles Machine*; ASTM C131/C131M-20: West Conshohocken, PA, USA, 2005.
36. American Concrete Institute. ACI Manual of Concrete Practice. In *Part 1: Materials and General Properties of Concrete*; American Concrete Institute: Farmington Hills, MI, USA, 2000.
37. Setareh, M.; Darvas, R. Reinforced Concrete Technology. In *Concrete Structures*; Springer: Cham, Switzerland, 2017; pp. 1–35. [CrossRef]
38. ASTM International. *Standard Test Method for Slump of Hydraulic Cement Concrete*; ASTMC143: West Conshohocken, PA, USA, 2000.
39. ASTM International. *Test Method for Compressive Strength of Cylindrical Concrete Specimens*; ASTMC39: West Conshohocken, PA, USA, 2014.

40. ASTM International. *Standard Test Method for Splitting Tensile Strength of Cylindrical Concrete Specimens*; ASTMC496: West Conshohocken, PA, USA, 2009.
41. Hasan, M.J.; Afroz, M.; Mahmud, H.M.I. An experimental investigation on the mechanical behavior of macro synthetic fibre reinforced concrete. *Int. J. Civ. Environ. Eng.* **2011**, *11*, 18–23.
42. Xu, L.; Li, B.; Ding, X.; Chi, Y.; Li, C.; Huang, B.; Shi, Y. Experimental Investigation on Damage Behavior of Polypropylene Fiber Reinforced Concrete under Compression. *Int. J. Concr. Struct. Mater.* **2018**, *12*, 68. [CrossRef]
43. Turner, R.P.; Kelly, C.A.; Fox, R.; Hopkins, B. Re-Formative Polymer Composites from Plastic Waste: Novel Infrastructural Product Application. *Recycling* **2018**, *3*, 54. [CrossRef]
44. Kamaruddin, M.A.; Abdullah, M.M.; Zawawi, M.H.; Zainol, M.R.R. Potential use of Plastic Waste as Construction Materials: Recent Progress and Future Prospect. *IOP Conf. Series Mater. Sci. Eng.* **2017**, *267*, 012011. [CrossRef]
45. Mazaheripour, H.; Ghanbarpour, S.; Mirmoradi, S.; Hosseinpour, I. The effect of polypropylene fibers on the properties of fresh and hardened lightweight self-compacting concrete. *Constr. Build. Mater.* **2011**, *25*, 351–358. [CrossRef]
46. Yin, S.; Tuladhar, R.; Shi, F.; Combe, M.; Collister, T.; Sivakugan, N. Use of macro plastic fibres in concrete: A review. *Constr. Build. Mater.* **2015**, *93*, 180–188. [CrossRef]
47. Aldahdooh, M.; Jamrah, A.; Alnuaimi, A.; Martini, M.; Ahmed, M. Influence of various plastics-waste aggregates on properties of normal concrete. *J. Build. Eng.* **2018**, *17*, 13–22. [CrossRef]
48. Ismail, Z.Z.; Al-Hashmi, E.A. Use of waste plastic in concrete mixture as aggregate replacement. *Waste Manag.* **2008**, *28*, 2041–2047. [CrossRef] [PubMed]
49. Albano, C.; Camacho, N.; Hernández, M.; Matheus, A.; Gutiérrez, A. Influence of content and particle size of waste pet bottles on concrete behavior at different w/c ratios. *Waste Manag.* **2009**, *29*, 2707–2716. [CrossRef] [PubMed]
50. Almeshal, I.; Tayeh, B.A.; Alyousef, R.; Alabduljabbar, H.; Mohamed, A.M. Eco-friendly concrete containing recycled plastic as partial replacement for sand. *J. Mater. Res. Technol.* **2020**, *9*, 1631–1643. [CrossRef]
51. Tasdemir, C.; Sengul, O.; Tasdemir, M.A. A comparative study on the thermal conductivities and mechanical properties of lightweight concretes. *Energy Build.* **2017**, *151*, 469–475. [CrossRef]
52. Alqahtani, F.K.; Ghataora, G.; Khan, M.I.; Dirar, S. Novel lightweight concrete containing manufactured plastic aggregate. *Constr. Build. Mater.* **2017**, *148*, 386–397. [CrossRef]
53. Mohammed, A.A.; Mohammed, I.I.; Mohammed, S.A. Some properties of concrete with plastic aggregate derived from shredded PVC sheets. *Constr. Build. Mater.* **2019**, *201*, 232–245. [CrossRef]
54. Awad, H.K. Influence of Cooling Methods on the Behavior of Reactive Powder Concrete Exposed to Fire Flame Effect. *Fibers* **2020**, *8*, 19. [CrossRef]
55. Neville, A.M. *Properties of Concrete*, 4th ed.; Longman Group: Essex, UK, 1995; p. 844.

Article

Mechanical and Market Study for Sand/Recycled-Plastic Cobbles in a Medium-Size Colombian City

Luz Adriana Sanchez-Echeverri [1,*], Nelson Javier Tovar-Perilla [2], Juana Gisella Suarez-Puentes [2], Jorge Enrique Bravo-Cervera [2] and Daniel Felipe Rojas-Parra [2]

[1] Facultad de Ciencias Naturales y Matemáticas, Universidad de Ibagué, Carrera 22 Calle 67, Ibagué 730002, Colombia

[2] Facultad de Ingeniería, Universidad de Ibagué, Carrera 22 Calle 67, Ibagué 730002, Colombia; nelson.tovar@unibague.edu.co (N.J.T.-P.); 2320161060@estudiantesunibague.edu.co (J.G.S.-P.); 2520161019@estudiantesunibague.edu.co (J.E.B.-C.); 2520162037@estudiantesunibague.edu.co (D.F.R.-P.)

* Correspondence: luz.sanchez@unibague.edu.co; Tel.: +57-8276-0010

Abstract: The need to satisfy the increasing demand for building materials and the challenge of reusing plastic to help improve the critical environmental crisis has led to the recycling of plastic waste, which is further exploited and transformed into new and creative materials for the construction industry. This study looked into the use of low-density recycled polyethylene (LDPE) to produce non-conventional plastic sand cobbles. LDPE waste was melted in order to obtain enough fluid consistency which was then mixed with sand in a 25/75 plastic-sand ratio respectively, such a mixture helped producing cobbles of 10 cm × 20 cm × 4 cm. Water absorption, weight, and density measurements were performed on both commercial and non-conventional plastic sand cobbles. Moreover, compression, bending, and wear resistance were also conducted as part of their mechanical characterization. Plastic sand cobbles showed lower water absorption and density values than commercial cobbles. The mechanical properties evaluated showed that plastic sand cobbles have a higher modulus of rupture and wear resistance than commercial cobbles. In addition, plastic sand cobbles meet the Colombian Technical Standard in lightweight traffic for pedestrians and vehicle, officially known as Norma Técnica Colombiana (NTC), with 25.5 MPa, 16.3 MPa, and 12 mm compression resistance, modulus of rupture and footprint length in wear resistance respectively. Finally, a market study was conducted to establish a factory to produce this type of cobbles in Ibague, Colombia. Not only the study showed positive financial indicators, which means that it is feasible running a factory to manufacture plastic sand cobbles in the city of Ibague, but it also concluded that nonconventional plastic sand cobbles could be explored to provide a comprehensive alternative to LDPE waste.

Keywords: materials; recycling; plastics; cobbles; lightweight traffic; pedestrian traffic

Citation: Sanchez-Echeverri, L.A.; Tovar-Perilla, N.J.; Suarez-Puentes, J.G.; Bravo-Cervera, J.E.; Rojas-Parra, D.F. Mechanical and Market Study for Sand/Recycled-Plastic Cobbles in a Medium-Size Colombian City. *Recycling* **2021**, 6, 17. https://doi.org/10.3390/recycling6010017

Received: 25 January 2021
Accepted: 26 February 2021
Published: 4 March 2021

1. Introduction

Plastic consumption and its latter disposal have become a problem due to the high volume of waste and the huge environmental impact they have, not only for the human population, but also for ecological systems [1–3]. Plastic is a versatile material with wide applications. However, it is a material that people do not consume correctly as there are no perceived dimensions on the environmental damage that its use entails [4,5].

Per capita plastic consumption continues to rise and remains high in high-income countries, despite obvious contributions to the global issue of plastic pollution [6]. In 2015, the World Bank concluded that if waste generation maintains the same dynamic without adequate actions to improve reuse, unsustainable use, and production, it will have become a health emergency issue in most countries. This is in addition to high greenhouse gas emissions by 2030.

Our planet is not capable of digesting the plastic waste generated daily and this will continue to happen. The best evidence is in the 4977 Mt of plastic waste accumulated in landfills or natural environments up until 2015 [7]. In Colombia, each person consumes an average of 2 kg of plastic per month, 24 kg per year, this means 1250 Mt per year in the country [8]. In 2018, about 31,500 t/day of plastic waste was disposed of in landfills, treatment plants, and waste cells—which means that of the 11.4 Mt of plastic waste in the year, only 17% was recycled [9]. The above situation shows what awaits humanity. Despite global efforts to solve these problems throughout recycling, plastic production has grown at a much higher rate. Plastic waste control cannot be achieved under the current circumstances. In Ibagué, Colombia, there are no waste management programs that could impact the population around recycling. For example, in 2018 only 4% of the city's waste was reused, because of the absence of recycling-companies in the region [10]. This poor level of recycling is constantly creating a crisis in the country's landfills due to the lack of waste disposal space. If viewed in a five-year perspective, the problem only tends to get worse, as it is estimated that 321 landfills in the country will reach their lifespan [11].

Different plastic recycling strategies have been proposed around the world: mechanical (classification, crushing, and cleaning), chemistry (pyrolysis, hydrolysis and glycolysis) and energy (plastic waste is used as fuel to produce electricity, steam, or heat) [12]. However, low-income countries have inadequate solid waste management with low waste collection rates; recycling infrastructure for these materials often does not exist in these countries. As a result, waste plastics have little or no value, resulting in uncontrolled disposal or leakage in the wild [6,13,14].

Another important sector that generates high environmental impact is the building and construction industry. For this reason, this industry has been interested in the development of more sustainable materials in recent years. To address this problem there has been development of cement-based materials with natural fibers using fewer polluting treatments or the replacement of cement with waste from different sources such LCD or e-waste [15–19]. However, cement, which is a high contaminant material, remains the main compound of these materials. In order to produce new materials with low cement content, there are different studies that have developed building materials from plastic waste [20–22], and in this way avoid tons of single-use plastic waste ending up in landfills.

Most of the plastic used in the manufacturing of these new building materials is polyethylene terephthalate-PET, which is plastic with the highest recycling percentage so far [23,24]. PET bottles have been used to make ecofriendly bricks; bottles are filled with sand or food wrappers [25,26]. Crushed PET has been used as an aggregate in the manufacturing of asphalt, concrete, and soils that show advantages such as weight reduction, corrosion resistance, and proper mechanical performance [27,28]. Although efforts to use plastic waste in the manufacturing of new building materials have shown good results, they still leave out alternatives for the use of a different type of plastic other than PET.

In order to create new materials from single-use plastic, in Africa the manufacture of construction blocks based on sand and plastics materials from Low Density Polyethylene-LDPE have been developed. LDPE is a type of single-use plastic that represents an environmental issue [29,30]. The manufacturing of these blocks has led examples of a community-driven waste management initiative that has had an impact on local communities and local waste management [31], not only as a job source but also for the implementation of these blocks in the poor condition's roads.

Kumi-Larbi Jnr, et al. (2018) have reported the manufacturing process to a laboratory scale of blocks using LDPE plastics only from water sachets [32]. However, there is no reported manufacturing process from other sources. This paper evaluates the mechanical performance of cobbles for pedestrian use based on plastic sand mix. Additionally, the market research and financial study are for the start-up of a manufacturing plant in Ibagué, Colombia of these non-conventional blocks with the aim to propose an alternative for an integrated waste management system.

2. Materials and Methods

2.1. Manufacturing of Plastic Sand Cobbles

LDPE is a thermoplastic that can be repeatedly molded when heated with a density in the range of 0.91–0.94 g/cm^3, with a melt temperature around 140 °C and water absorption <0.01%. LDPE from different plastics waste was supplied by a local recycled plastic bags manufacturer, LDPE was obtained crushed and pelletized in a 3 mm diameter to facilitate its melt; sand was supplied by local building materials enterprise (HOMECENTER SODIMAC corona). Figure 1 shows the raw materials used to manufacture plastic sand cobbles (see Figure 1).

Figure 1. Sand and plastic pellets used in the manufacturing of plastic sand cobbles.

Commercially available silica sand was used as an inert filler with a particle density of 2.65 g/cm^3 with 1.1% of fine aggregate and a sieve analysis (INV E 123-13) shows in Figure 2.

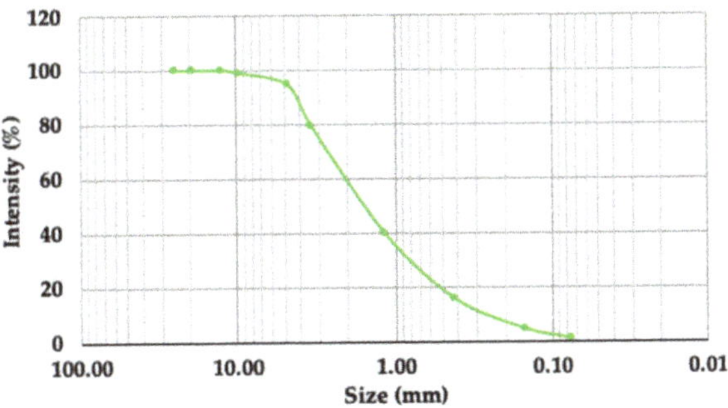

Figure 2. Sieve analysis for commercially sand used in the sand/plastic cobbles.

The sand was previously baked dry for 24 h and sifted with particle sizes of <1.00 mm diameter as is suggest in previous results [32].

For LDPE plastic sand cobbles, a mixture with a ratio of 25/75 respectively was used due to previously reported studies; for this work 1600 g of sand and 400 g of plastic pellets were employed for the cobbles manufacturing. The process flow diagram, as well as the plastic sand block obtained, are shown in Figure 3. The plastic pellets are melted in an oven at 180 °C $\pm$ 5 °C; once the plastic had the proper consistency the sand was added and mixed continuously until obtained a homogeneous mix. The mixture was placed in a 10 cm × 20 cm × 4 cm casting mold. Two hours after the sample was demoulding, and it was kept at room temperature before the analyses; a total of 18 cobbles were manufactured to carried out the different tests; 5 for each mechanical testing (MOR and Compressive strength), 5 for water absorption and only 3 for wear resistance due the test was developed in an external laboratory, the same quantity of commercial cobbles was tested. The physical characteristics (dimensions, density and weight) of both cobbles sand/plastic and commercial were determinate in the 18 samples because they are nondestructive testing.

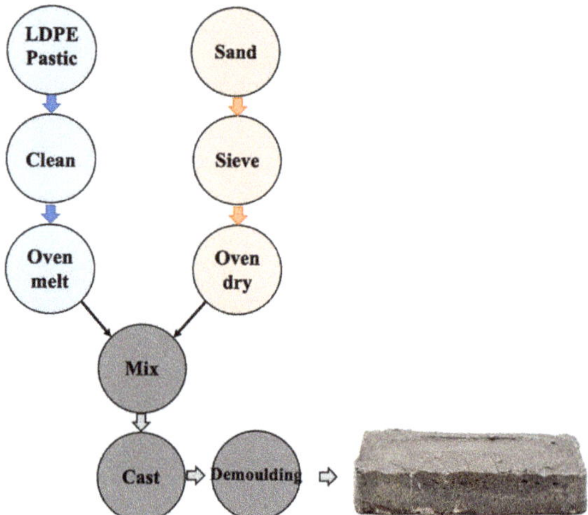

Figure 3. Flow diagram for plastic sand cobbles manufacturing.

2.2. Plastic Sand Cobbles Characterization

LDPE plastic sand blocks were characterized and compared with commercial blocks of pedestrian use. The water absorption of samples was determined after 24 h of immersion in distilled water at room temperature, as described in ASTM D570. Density and dimensions were calculated by direct measurements as is indicated in the Colombian Technical Standard, known officially as Norma Técnica Colombiana (NTC 2017-Pavement concrete cobblestones) [33]. Water absorption for plastic sand and commercial cobbles was calculated following equation

$$\text{Water absorption } (\%) = [\frac{Ww - Wd}{Ww}] * 100\%, \tag{1}$$

where:

Ww = Wet weight of specimen [g]
Wd = Dry weight of specimen [g]

Tests were performed for the mechanical characterization of the modulus of rupture (MOR) and compression resistance of both the plastic sand and commercial cobbles. A three-point load bending test with a span of 200 mm and a speed of 10 mm/min was performed in accordance with Colombian Technical Standard [33] on a JJ Lloyd traction test machine. Dry specimens were tested in a room in a controlled environment at 23 °C $\pm$ 2 °C and a relative humidity of 50 $\pm$ to 5%. Equation (2) determines the modulus of rupture and Figure 4 shows a detail of variables involved:

$$\text{MOR} = \frac{3L_{max}(l_i - 20)}{[(w_r + w_i)\, e^2]}, \tag{2}$$

where:

L_{max} = Maximum load [N]
l_i = Length of inscribed rectangle [mm]
w_r = Real width [mm]
w_i = Inscribed rectangle width [mm]

e = Thickness [mm]

Figure 4. Representation of variables involved in Modulus of Rupture.

Finally, the wear resistance of plastic sand and commercial cobbles was determined on the basis of Colombian Technical Standard, (NTC) 5147 2002 (This standard is the determined test method for measuring resistance to abrasion of materials for floors and pavements; this test is used to determine the abrasion resistance of floor and pavement materials by sand and width of disc). The cobbles were subjected to wear by means of abrasion exerted, under controlled conditions, by a flow of sand that passes tangentially between that surface and the side face of a metal disc, which exerts pressure against it. This generates a footprint, in the shape of the curved surface of the metal disc (Figure 5).

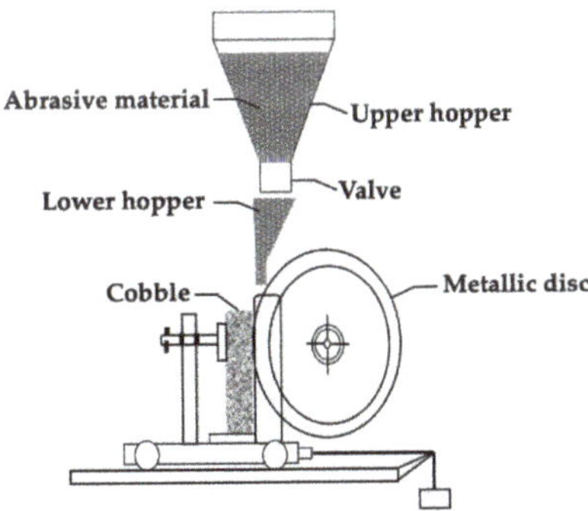

Figure 5. Sketch of wear resistance test for commercial and sand/plastic cobbles.

After test, the footprint generated was measured as follows Equation (3)

$$f_l = AB + (200 - V_c),\qquad(3)$$

where:

    f_l = Length of footprint [mm]
    AB = Length of footprint in the center point [mm]
    V_c = Calibration value [mm]

The resulting length is inversely proportional to the abrasive wear resistance, which the sample possesses. The test was carried out in 3 samples in order to obtain an average abrasion resistance value.

2.3. Market Research—Case of Study: Ibagué, Colombia

This study encompasses all the necessary and relevant information to determine the feasibility of investment in this type of project. The feasibility study covers four main areas: Market Study, Technical Study, Organizational Study, Financial Study [34].

Market Study: The market study will indicate the acceptability of the cobbles. The target population was companies and entities based in the city of Ibagué with economic activity characterized by the purchase or marketing of cobblestones. The population was segmented in accordance with the International Uniform Industrial Classification (IUIC) of Higher Education Institutions: agro-tourism farms and companies dedicated to the construction of pedestrian roads and hardware stores [35]. Equation (4) with a confidence level of 10% was used to determine the sample size.

$$n = \frac{Z^2_\alpha \cdot N \cdot p \cdot q}{i^2(N-1) + Z^2_\alpha \cdot p \cdot q},\tag{4}$$

where:

n: Sample size

Z: Value corresponding to Gauss distribution for significance level α (for 10% significance takes the value of 1.64)

N: Population size (2246 companies)

p: Expected prevalence of the parameter to be evaluated, if $p = 0.5$ is unknown (q: $1-p$)

i: Error expected to be made (10% $\times$ 0.1)

Interviews were conducted with the people in charge of the sales or purchase of building materials in order to know the level of acceptance of plastic sand cobblestones and the demand of the product in the market.

Financial Study: Compiles all financial information for the implementation of such projects and converts it to monetary terms to look at the approval or rejection of the proposal.

3. Results and Discussion

3.1. Physical Characteristics of Cobbles

The physical characteristics of plastic sand and commercial cobbles after manufacturing are shown in Table 1.

Table 1. Physical characteristics of plastic sand and commercial cobbles.

Physics Characteristics	Plastic/Sand Cobble	Commercial Cobble
Length [cm]	19.20 [a] 0.87 [b] 0.21 [c]	20.01 [a] 0.11 [b] 0.03 [c]
Width [cm]	9.50 [a] 0.33 [b] 0.08 [c]	10.00 [a] 0.09 [b] 0.02 [c]
Thickness [cm]	4.00 [a] 0.27 [b] 0.06 [c]	5.90 [a] 0.21 [b] 0.05 [c]
Weight [g]	1232.60 [a] 118.02 [b] 27.82 [c]	2438.94 [a] 121.87 [b] 28.72 [c]
Density [g/cm^3]	1.59 [a] 0.10 [b] 0.02 [c]	2.08 [a] 0.06 [b] 0.01 [c]
Water absorption [%]	2.56 [a] 0.78 [b] 0.35 [c]	6.30 [a] 1.50 [b] 0.67 [c]

[a] Mean, [b] Standard deviation and [c] Typical Error.

According to the Colombian Technical Standard for pedestrian and light vehicle traffic [33], the dimensions of the cobblestones (concrete cobblestones for pavement) must be 100 mm < length $\leq$ 250 mm, width $\geq$ 100 mm and 40 mm < thickness $\leq$ 60 mm. According to dimensions, the results of cobbles produced with plastic waste and sand samples obtained the necessary dimensions to be marketed as pedestrian or light traffic vehicle cobbles. This is a wide potential market. Based on mass and density results, the addition of plastic in the cobbles reduces the weight and density of the sample by 49.46% and 23.56% respectively. This implies the production of lighter materials; it has been shown to be an advantage in the search for new building materials [36]. Lightweight materials reduce the cost of transportation and marketing. In addition, conventional brick masonry was ranked highest in terms of CO_2 emissions, while sand lime masonry which is light in nature, was ranked as the lowest [37]. Finally, water absorption, which is the capacity that water has to penetrate into samples, is closely related to porous distribution and porosity of materials. This is an important indicator of long-term mechanical properties. According to the results of Table 1, the plastic sand cobbles reduce water absorption by approximately 60%. This reduction is due to the hydrophilic nature of plastic. Plastic sand cobbles showed an improvement in resistance to water absorption, and this behavior is based on previous results of bricks containing crushed glass and polypropylene granules [38,39]. According to water absorption results and based on the Colombian Technical Standard (NTC) 3829

(Clay Cobblestone for Pedestrian Transit and Light Vehicle) plastic sand cobbles could be classified as Type I. This means that cobbles can be exposed to high abrasion environments.

3.2. Mechanical Characteristics of Cobbles

The modulus of rupture (MOR) and the compression resistance of plastic sand and commercial cobbles are shown in Table 2; for MOR values, the thickness of the cobbles was measured at four different points and the mean value was used. MOR is the most important mechanical feature for this type of materials used for pedestrians and light traffic because they work predominantly in bending. The plastic sand cobbles rupture module is 1.5 times larger than commercial. However, the dispersion obtained for plastic sand cobbles is greater than its commercial counterpart due to its artisanal manufacturing process, which made differences in the final material. Different studies have shown an increase in the modulus of rupture of building materials with addition of plastic, which agrees with the results found here [39,40]. On the other hand, compression resistance is higher for commercial sand cobbles than for the plastics ones.

Table 2. Compression Modulus of rupture of commercial and plastic sand cobbles.

Type of Cobble	Compression Resistance (MPa)	Modulus of Rupture (MPa)
Commercial cobbles	30.60 [a] 3.40 [b] 1.52 [c]	6.03 [a] 0.40 [b] 0.17 [c]
Plastic sand cobbles	25.49 [a] 1.70 [b] 0.76 [c]	16.28 [a] 2.51 [b] 1.12 [c]

[a] Mean, [b] Standard deviation and [c] Typical Error.

Wear resistance is one of the parameters for measuring the durability of materials exposed to different environments. The length of the footprint after the wear test was 21.10 mm $\pm$ 0.66 mm and 12.43 $\pm$ 0.80 mm for commercial cobbles and plastic sand cobbles, respectively. Arango-Londoño (2006) determined an average footprint length of 21.7 mm for concrete cobbles with a maximum value of 23 mm [41]. These values match the commercial cobbles studied and are in accordance with the recommended values for various standards [42]. Wear resistance has no association with the other mechanical characteristics analyzed because it behaves as a separate variable. Therefore, it must be specified as a product quality control test independently, without being indirectly controlled from other variables. For regulatory purposes, it is recommended to specify a maximum tread length in the abrasion resistance test of 23 mm [41].

In accordance with the requirements of the Colombian Government for Pedestrian Areas and Light Traffic Areas, the standard establishes minimum values of 10 MPa of MOR tests under dry conditions for vehicle traffic, 6% and 8% water absorption for pedestrian and vehicle traffic respectively, 55 MPa and 69 MPa of compression resistance for pedestrian and vehicle traffic respectively, and 0.11 of the abrasion rates for both uses of pedestrian and vehicle traffic. According to this standard, the results obtained in the physical and mechanical characterization of the plastic and sand mixture allow to obtain suitable materials for these applications. Moreover, they allow the reduction of plastic waste and they could contribute to reduce the water crisis due the manufacture process do not consume this resource.

3.3. Market and Financial Study on Ibagué

Table 3 shows the target population and sample size calculated according to Equation (1); the results were obtained based on information of companies registered to the Ibague Chamber of Commerce as of 31 May 2018.

In order to know the demand for these types of products, purchase frequency was investigated, showing that the daily and weekly frequencies of purchase were more recurring (Figure 6). In addition, the most representative entities in this frequency of purchase are retail hardware stores which represent 69.2% of the target market, and wholesale stores representing 15.4% of the target market. With these results, it was identified that it is necessary to adopt a production process that can meet a market with daily and weekly demand.

Table 3. Target population and sample size calculated for the market study of the start-up of a nonconventional plastic sand cobbles plant in the city of Ibague—Colombia.

Activity	CIIU Code	Total	Sample
Farms–Rural accommodation	I5514	77	2
Non-permanent tourist housing, excluding hourly accommodation	I5519	46	1
	G4752	1554	46
Retail trade in hardware items	G4663	346	10
Wholesale trade in building materials and hardware items	F4390	178	5
	P8543	6	1
Other specialized activities for the construction	P8523	39	1
Total		**2246**	**66**

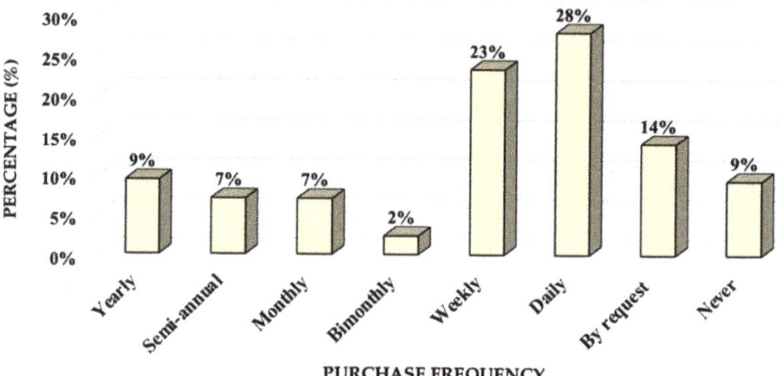

Figure 6. Frequency of purchase of cobbles for pedestrian use or light vehicle traffic.

Once demand behavior is known, it is necessary to determine what type of companies the population buy the product from. As shown in Figure 7, 56% of the target population prefers to buy bricks through a distributor, 26% directly from the manufacturer, 13% in a hardware store, and the remaining 5% from a specialty company.

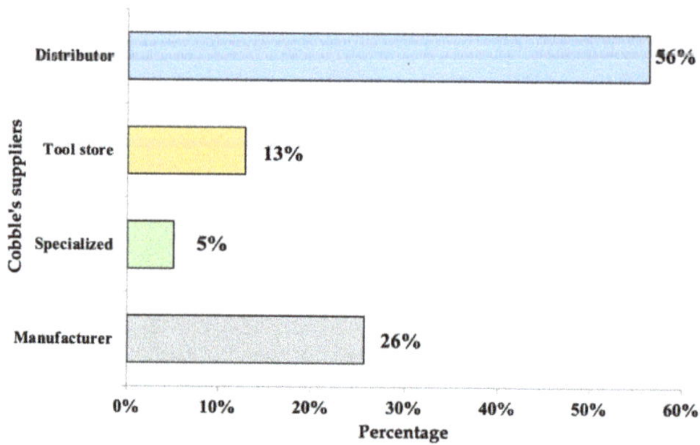

Figure 7. Suppliers of cobbles for pedestrians use in entities interviewed in Ibagué.

According to the above results, companies wishing to manufacture plastic sand bricks should consider within their positioning strategies both direct (manufacturer) and indirect distribution channels (distributors), because results show an increasing inclination of the buyers to purchase through distributors.

This market research also sought to find out what knowledge different entities have about the environmental impact of plastic waste, what actions they take to protect the environment, and whether their business policies consider aspects related to the conservation of the environment. It was found that 58% of the companies do not take any action to reduce plastic waste generation, although 86% of them are aware of the impact it has. On the other hand, the remaining 42% of the market does take steps to reduce it, but most rely on basic recycling such as paper and cardboard recycling. The sector of companies that market products for pedestrian construction does not yet understand the serious inherent impact that exists on the use and subsequent disposal of plastics, which can be caused by the current little or no legislation in Colombia on the management of this waste.

The satisfaction of both, products and suppliers were evaluated (Figure 8). You can see that the market is a little more satisfied with its suppliers than with the products.

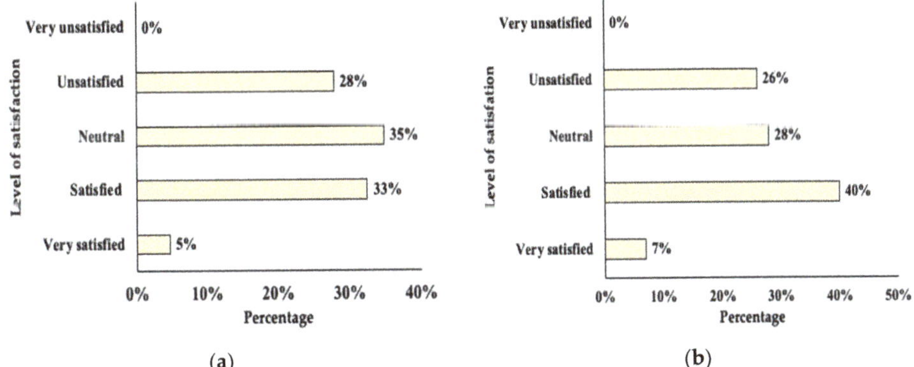

Figure 8. Perception of satisfaction (**a**) with products available in the market and (**b**) with suppliers of the market.

Product dissatisfaction focuses on durability and price. Plastic sand cobbles have a competitive advantage in offering a cobble with a longer shelf life than the materials currently used because of the resistance offered by plastic. However, regarding the price of the product, it should be taken into account whether the market is willing to pay a higher price. Supplier dissatisfaction and neutral opinion remain more than half of the market (54%). About 28% are dissatisfied due to supplier noncompliance and the remaining 22% because of poor quality of the products they offer. Finally, the degree of acceptance of such products on the market was evaded. The results show that 100% of the people surveyed are willing to buy or market unconventional cobbles made of recycled plastic. However, it should be taken into account that only 28% of the market is dissatisfied, so this percentage will be the limiting factor for the competition. The main characteristics by which customers would buy the product are the decrease in environmental impacts when using recycled plastic and the high durability of the product. The survey also showed that 67% of respondents would be willing to pay a higher price for an ecological cobble.

It is important to note that there is a relationship between the lack of knowledge about the environmental damage caused by plastics and whether potential customers would be willing to pay the price for the new plastic sand cobbles. Companies that have no knowledge on environmental issues are not willing to pay a higher price compared to the ones that are accustomed to paying these prices. An estimate of the demand for the potential customers measured in the number of cobbles showed that for the market of the city of Ibagué, 2500 cobbles are projected annually. Bearing in mind the limitations that arise due to market competition (28%), a demand of 1500 cobbles is estimated.

To perform the financial analysis, a cash flow was determined from raw material costs, personnel, indirect manufacturing costs, overhead and financial costs, and investments required for the start-up of the project (Table 4); the values presented in Table 4 are based in the study titled, "Study of the feasibility for the production of non-conventional cobbles using recycled Low-density polyethylene in the city of Ibagué", which is a thesis that was part of the construction of this work [43]. The cost of production was also set at USD 0.59 and with a contribution margin of approximately 25% a sale price of USD 0.74 was set.

Table 4. Five-year projection cash flow for a plastic sand cobbles start-up manufacturing company (in miles of USD).

	Year 0	Year 1	Year 2	Year 3	Year 4	Year 5
Income						
Sales Revenue		438.	454.7	943.2	978.2	1014.7
Total entries		438.4	454.7	943.2	978.2	1014.7
Outcome						
Raw materials		63.1	64.8	133.1	136.6	140.2
Payroll		56.2	58.8	94.6	98.8	109.5
CIF		50.8	52.4	91.8	95.8	100.7
Overhead		32.9	34.0	43.2	45.1	51.6
Financial expenses		9.1	7.7	6.1	4.3	2.3
Depreciation		6.8	6.8	6.8	6.8	6.8
Amortization		0.3	0.3	0.3	0.3	0.3
Total, outcome		219.3	224.8	376.4	388.3	412.2
Profit before tax		219.0	229.9	566.8	590.0	602.4
Income tax		70.1	73.6	181.4	188.8	192.8
Net income		148.9	156.3	385.4	401.2	409.6
Depreciation		6.7	6.7	7.4	7.4	7.6
Amortization		0.3	0.3	0.3	0.3	0.3
Investments						
Fixed assets investments	−76.4			4.1		1.1
Investment in pre-operating	−1.5					
Working capital investment	−32.3					
Capital credit		12.2	13.6	15.2	17.1	19.1
Credit	77.2					
Residual value						83.4
Available balance	**−33.1**	**143.8**	**149.8**	**373.7**	**391.8**	**480.7**

With the values in the table, the net present value (NPV) and the cost-benefit ratio (CBR) were calculated according to Equations (5) and (6):

$$\text{NPV} = [Sd_{year\ 1} * (1+td)^{-1}] + [Sd_{year\ 2} * (1+td)^{-2}] + [Sd_{year\ 3} * (1+td)^{-3}] + [Sd_{year\ 4} * (1+td)^{-4}] + [Sd_{year\ 5} * (1+td)^{-5}] - inicial\ investment, \quad (5)$$

where:

Sd: Cash flow available balance from Investor (Sum of net profit, amortization and provisions)

td: Discount rate (%) (Sum of liability rate and country-risk rate)

$$CBR = \frac{NPV_e}{NPV_s}, \quad (6)$$

where:

NPV_e: Net present value of revenues

NPV_s: Net present value of expenses

The NPV is approximately \$987 USD. This value represents the net cash profit generated during the execution of the project, after the initial investment is covered. Likewise, the cost benefit rate yielded the value of 1.58 in profits for each dollar invested; these indicators show the feasibility of the project as a NPV > 0 and a CBR > 1 were obtained.

4. Conclusions

This study has looked into the use of plastic waste (LDPE) and sand to manufacture cobbles. From the results of the tests carried out for physical properties, plastic sand cobbles showed a decrease in the water absorption rate and its weight by around 50% compared to commercial cobbles from 12.4% to 6.2% and 2.4 kg to 1.2 kg, respectively. Plastic sand cobbles have a compression resistance value of 25.5 MPa which is higher than recommended in the standard, although this value is less than commercial which is 30.6 MPa. Plastic sand cobbles also showed a higher modulus of rupture of 6.03 MPa compared to 16.3 MPa for commercial cobbles which is about three times higher. Both physical and mechanical properties of plastic sand cobbles meet the requirements for Colombian standards. Therefore, this type of cobbles can be installed in areas with light pedestrian and vehicular traffic.

According to the market study conducted to establish a factory for the implementation of processing and transformation of low-density polyethylene (LDPE), the ultimate aim of reducing the problem of plastic waste in the city of Ibague is viable. Additionally, creating an alternative for comprehensive recycling is observed because plastic sand cobbles meet physical and mechanical requirements. In the same way these cobbles meet the current demands in a more sustainable approach. The financial indicators determined that the initial investment would be maximized according to costs, the financial option, and working capital and sales. In addition, the cost profit rate showed a positive value indicating that the project generates surplus funds according to financial projections.

Author Contributions: Conceptualization, L.A.S.-E. and N.J.T.-P.; methodology, L.A.S.-E. and N.J.T.-P.; formal analysis L.A.S.-E. and N.J.T.-P.; investigation, J.E.B.-C., D.F.R.-P., J.G.S.-P.; data curation, J.E.B.-C., D.F.R.-P., J.G.S.-P.; writing—original draft preparation, J.E.B.-C., D.F.R.-P., J.G.S.-P.; writing—review and editing, L.A.S.-E. and N.J.T.-P.; supervision, L.A.S.-E. and N.J.T.-P. All authors have read and agreed to the published version of the manuscript.

Funding: This research received no external funding.

Informed Consent Statement: Not applicable.

Data Availability Statement: New data were created or analyzed in this study. Data will be shared upon request and consideration of the authors.

Acknowledgments: The authors appreciate the time enterprises interviewed spent and the technical support in the laboratory procedures provided by laboratory staff at Universidad de Ibague.

Conflicts of Interest: The authors declare no conflict of interest.

References

1. Cordier, M.; Uehara, T. How much innovation is needed to protect the ocean from plastic contamination? *Sci. Total Environ.* **2019**, *670*, 789–799. [CrossRef] [PubMed]
2. Rai, P.K.; Lee, J.; Brown, R.J.C.; Kim, K.H. Environmental fate, ecotoxicity biomarkers, and potential health effects of micro- and nano-scale plastic contamination. *J. Hazard. Mater.* **2021**, *403*, 123910. [CrossRef] [PubMed]
3. Ribeiro-Brasil, D.R.G.; Torres, N.R.; Picanço, A.B.; Sousa, D.S.; Ribeiro, V.S.; Brasil, L.S.; de Assis Montag, L.F. Contamination of stream fish by plastic waste in the Brazilian Amazon. *Environ. Pollut.* **2020**, *266*, 115241. [CrossRef]
4. Beaumont, N.J.; Aanesen, M.; Austen, M.C.; Börger, T.; Clark, J.R.; Cole, M.; Hooper, T.; Lindeque, P.K.; Pascoe, C.; Wyles, K.J. Global ecological, social and economic impacts of marine plastic. *Mar. Pollut. Bull.* **2019**, *142*, 189–195. [CrossRef] [PubMed]
5. Rhein, S.; Schmid, M. Consumers' awareness of plastic packaging: More than just environmental concerns. *Resour. Conserv. Recycl.* **2020**, *162*, 105063. [CrossRef]
6. Barnes, S.J. Out of sight, out of mind: Plastic waste exports, psychological distance and consumer plastic purchasing. *Glob. Environ. Chang.* **2019**, *58*, 101943. [CrossRef]
7. Geyer, R.; Jambeck, J.R.; Law, K.L. Production, use, and fate of all plastics ever made. *Sci. Adv.* **2017**, *3*, 25–29. [CrossRef]

8. El Heraldo; Alianza Unicosta. Planeta Plástico 2019. Available online: https://www.elheraldo.co/barranquilla/planeta-plastico-618648 (accessed on 11 November 2020).
9. Superintendencia de Servicios Públicos Domiciliarios Disposición Final de Residuos Sólidos-Informe Nacional 2018. 2019; p. 97. Available online: https://www.superservicios.gov.co/sites/default/archivos/Publicaciones/Publicaciones/2020/Ene/informe_nacional_disposicion_final_2019_1.pdf (accessed on 11 October 2020).
10. Ecos del Combeima Ibagué Seguirá Adelante con el Programa de Reciclaje y Recolección de Basuras 2018. Available online: https://www.ecosdelcombeima.com/ibague/nota-125575-ibague-seguira-adelante-con-el-programa-de-reciclaje-y-recoleccion-de-basuras (accessed on 5 October 2020).
11. Urrutia, C. Qué Retos Enfrenta el Reciclaje y Cómo Podemos Ayudar los Individuos? 2018. Available online: https://sostenibilidad.semana.com/impacto/multimedia/que-retos-enfrenta-el-reciclaje-y-como-podemos-ayudar-los-individuos/42399#:~:text=Colombia%20genera%20anualmente%20más%20de,la%20preservación%20del%20medio%20ambiente (accessed on 8 October 2020).
12. Franco-Urquiza, E.; Ferrado, H.; Luis, D.; Maspoch, M.L. Reciclado mecánico de residuos plásticos. Caso práctico: Poliestireno de alto impacto para la fabricación de componentes de TV. *Afinidad* **2016**, *73*, 575.
13. Goli, V.S.N.S.; Mohammad, A.; Singh, D.N. Application of Municipal Plastic Waste as a Manmade Neo-construction Material: Issues & Wayforward. *Resour. Conserv. Recycl.* **2020**, *161*, 105008. [CrossRef]
14. Parveen, N. UK's Plastic Waste May Be Dumped Overseas Instead of Recycled. *Guard* **2018**. Available online: https://www.theguardian.com/environment/2018/jul/23/uks-plastic-waste-may-be-dumped-overseas-instead-of-recycled (accessed on 3 March 2021).
15. Da Costa, V.; Santos, S.F.; Mármol, G.; Da Silva, A.A.; Savastano, H. Potential of bamboo organosolv pulp as a reinforcing element in fiber–cement materials. *Constr. Build. Mater.* **2014**, *72*, 65–71. [CrossRef]
16. Sanchez-Echeverri, L.A.; Medina-Perilla, J.A.; Ganjian, E. Nonconventional Ca(OH)$_2$ treatment of bamboo for the reinforcement of cement composites. *Materials* **2020**, *13*, 1892. [CrossRef]
17. Sanchez-Echeverri, L.A.; Ganjian, E.; Medina-Perilla, J.A.; Quintana, G.C.; Sanchez-Toro, J.H.; Tyrer, M. Mechanical refining combined with chemical treatment for the processing of Bamboo fibres to produce efficient cement composites. *Constr. Build. Mater.* **2021**, *269*, 121232. [CrossRef]
18. Kim, S.; Hanif, A.; Jang, I.Y. Incorporating Liquid Crystal Display (LCD) Glass Waste as Supplementary Cementing Material (SCM) in Cement Mortars—Rationale Based on Hydration, Durability, and Pore Characteristics. *Materials* **2018**, *11*, 2538. [CrossRef]
19. Suchorab, Z.; Franus, M.; Barnat-hunek, D. Properties of Fibrous Concrete Made with Plastic Optical Fibers from E-Waste. *Materials* **2020**, *13*, 2414. [CrossRef]
20. Appiah, J.K.; Berko-Boateng, V.N.; Tagbor, T.A. Use of waste plastic materials for road construction in Ghana. *Case Stud. Constr. Mater.* **2017**, *6*, 1–7. [CrossRef]
21. Arulrajah, A.; Yaghoubi, E.; Wong, Y.C.; Horpibulsuk, S. Recycled plastic granules and demolition wastes as construction materials: Resilient moduli and strength characteristics. *Constr. Build. Mater.* **2017**, *147*, 639–647. [CrossRef]
22. Santos, J.; Pham, A.; Stasinopoulos, P.; Giustozzi, F. Recycling waste plastics in roads: A life-cycle assessment study using primary data. *Sci. Total Environ.* **2021**, *751*, 141842. [CrossRef] [PubMed]
23. Lonca, G.; Lesage, P.; Majeau-Bettez, G.; Bernard, S.; Margni, M. Assessing scaling effects of circular economy strategies: A case study on plastic bottle closed-loop recycling in the USA PET market. *Resour. Conserv. Recycl.* **2020**, *162*, 105013. [CrossRef]
24. Majumdar, A.; Shukla, S.; Singh, A.A.; Arora, S. Circular fashion: Properties of fabrics made from mechanically recycled poly-ethylene terephthalate (PET) bottles. *Resour. Conserv. Recycl.* **2020**, *161*, 104915. [CrossRef]
25. Lenkiewicz, Z.; Webster, M. Making Waste Work: A Toolkit How to Turn Mixed Plastic Waste and Bottles into Ecobricks a Step-By-Step Guide. 2017. Available online: https://wasteaid.org/wp-content/uploads/2017/10/9-How-to-turn-mixed-plastic-waste-and-bottles-into-ecobricks-v1-mobile.pdf (accessed on 3 March 2021).
26. Mansour, A.M.H.; Ali, S.A. Reusing waste plastic bottles as an alternative sustainable building material. *Energy Sustain. Dev.* **2015**, *24*, 79–85. [CrossRef]
27. Gu, L.; Ozbakkaloglu, T. Use of recycled plastics in concrete: A critical review. *Waste Manag.* **2016**, *51*, 19–42. [CrossRef]
28. Vasudevan, R.; Ramalinga Chandra Sekar, A.; Sundarakannan, B.; Velkennedy, R. A technique to dispose waste plastics in an ecofriendly way-Application in construction of flexible pavements. *Constr. Build. Mater.* **2012**, *28*, 311–320. [CrossRef]
29. Marsh, K.; Bugusu, B. Food packaging-Roles, materials, and environmental issues: Scientific status summary. *J. Food Sci.* **2007**, *72*, R39–R55. [CrossRef] [PubMed]
30. Abraham, J.; Ghosh, E.; Mukherjee, P.; Ganjendiran, A. Microbial Degradation of Low Density Polyethylene. *Environ. Prog. Sustain. Energy* **2016**, *36*, 147–154. [CrossRef]
31. Wilson, D.C.; Webster, M. Building capacity for community waste management in low- and middle-income countries. *Waste Manag. Res.* **2018**, *36*, 1–2. [CrossRef]
32. Kumi-Larbi, A.; Yunana, D.; Kamsouloum, P.; Webster, M.; Wilson, D.C.; Cheeseman, C. Recycling waste plastics in developing countries: Use of low-density polyethylene water sachets to form plastic bonded sand blocks. *Waste Manag.* **2018**, *80*, 112–118. [CrossRef]
33. NTC 2017 Normas Técnica Colombiana Ntc 2017 Adoquines De Concreto Para Pavimento-Metroblock. Available online: https://metroblock.com.co/norma-tecnica-colombiana-ntc-2017/ (accessed on 3 March 2021).

34. Sapag Chain, N.; Sapag Chain, R. *Preparación y Evaluación de Proyectos*; Quinta; McGraw-Hill Interamericana: Mexico City, Mexico, 2008; ISBN 978-956-278-206-7.

35. DANE Clasificación Industrial Internacional Uniforme De Todas Las Actividades Económicas. 2012; p. 496. Available online: https://www.dane.gov.co/files/sen/nomenclatura/ciiu/CIIURev31AC.pdf (accessed on 3 March 2021).

36. Shah, S.N.; Mo, K.H.; Yap, S.P.; Yang, J.; Ling, T.C. Lightweight foamed concrete as a promising avenue for incorporating waste materials: A review. *Resour. Conserv. Recycl.* **2021**, *164*, 105103. [CrossRef]

37. Rama Jyosyula, S.K.; Surana, S.; Raju, S. Role of lightweight materials of construction on carbon dioxide emission of a reinforced concrete building. *Mater. Today Proc.* **2020**, *27*, 984–990. [CrossRef]

38. Akinyele, J.O.; Igba, U.T.; Ayorinde, T.O.; Jimoh, P.O. Structural efficiency of burnt clay bricks containing waste crushed glass and polypropylene granules. *Case Stud. Constr. Mater.* **2020**, *13*, e00404. [CrossRef]

39. Akinyele, J.O.; Igba, U.T.; Adigun, B.G. Effect of waste PET on the structural properties of burnt bricks. *Sci. Afr.* **2020**, *7*, e00301. [CrossRef]

40. Farrapo, C.; Soares, C.; Pereira, T.G.T.; Tonoli, G.H.D.; Junior, H.S.; Mendes, R.F. Cellulose associated with pet bottle waste in cement based composites. *Mater. Res.* **2017**, *20*, 1380–1387. [CrossRef]

41. Arango-Londoño, J.F. Adoquines en concreto: Propiedades físico-mecánicas y sus correlaciones. *TecnoLógicas* **2006**, *16*, 121–137. [CrossRef]

42. CEN BS EN 12467:2004 Fibre-Cement Flat Sheets—Product Specification and Test Methods. 2004, p. 52. Available online: https://shop.bsigroup.com/ProductDetail/?pid=000000000030131889 (accessed on 3 March 2021).

43. Suarez-Puentes, J.G. Estudio de la Factibilidad para la Producción de Adoquines no Convencionales a Partir de la Reutilización del Polietileno de baja Densidad en la Ciudad de Ibagué. Bachelor's Thesis, Universidad de Ibagué, Ibagué, Colombia, 2020. Available online: https://hdl.handle.net/20.500.12313/2314 (accessed on 1 February 2021).

 recycling

Article

Mechanical Property Assessment of Interlocking Plastic Pavers Manufactured from Electronic Industry Waste in Brazil

Luiz Tadeu Gabriel [1,2], Rodrigo Fernando Bianchi [1,2] and Américo T. Bernardes [1,*]

1 Programa Ciências: Física de Materiais, Departamento de Física, Instituto de Ciências Exatas e Biológicas, Universidade Federal de Ouro Preto, Ouro Preto 35400-000, Brazil; luiztadeugabriel@yahoo.com.br (L.T.G.); bianchi@ufop.edu.br (R.F.B.)
2 Laboratório de Polímeros e de Propriedades Eletrônicas de Materiais, Departamento de Física, Instituto de Ciências Exatas e Biológicas, Universidade Federal de Ouro Preto, Ouro Preto 35400-000, Brazil
* Correspondence: atb@ufop.edu.br; Tel.: +55-31-99100-4664

Abstract: The estimated production of world electronic waste until 2017 is approximately 6 Gt. Despite this enormous problem, there are no clear regulations regarding the orientation for disposal or treatment of this type of residuals in many countries. There is a federal public policy in Brazil that supports a network of Computer Reconditioning Centers—CRCs. These CRCs train young people and recover or recycle electronic equipment. Through this work, CRCs produce interlocking plastic pavers for application on pavements from recycled electronic industry waste. This article presents the characterization of these interlocking paver's mechanical properties when applied on the pavement. This characterization is a necessary step to show the effectiveness of this product. We show that the plastic pavers behave similarly to the artifacts manufactured in concrete, thus creating commercial opportunities for this initiative, and contributing to the Brazilian Solid Waste Policy.

Keywords: electric–electronic waste; interlock floor; mechanical resistance; polymers recycling

Citation: Gabriel, L.T.; Bianchi, R.F.; Bernardes, A.T. Mechanical Property Assessment of Interlocking Plastic Pavers Manufactured from Electronic Industry Waste in Brazil. *Recycling* **2021**, *6*, 15. https://doi.org/10.3390/recycling6010015

Received: 9 November 2020
Accepted: 16 February 2021
Published: 25 February 2021

Publisher's Note: MDPI stays neutral with regard to jurisdictional claims in published maps and institutional affiliations.

1. Introduction

1.1. Generation of Electronic Waste in the World and Opportunities for the Circular Economy

It has been estimated that until 2017, 8.3 Gt of virgin polymers have been produced, about 75% became waste. Of this waste, in 2015, approximately 9% were recycled, 12% were incinerated, and 79% were accumulated. By 2050, primary waste generation is projected to reach about 25 Gt, with a tendency to stabilize waste disposal to approximately 13 Gt, an increase in incineration to approximately 12 Gt, and an increase in recycling to around 9 Gt [1].

Electronic waste is emerging as a new environmental challenge of the 21st century due to the rapid escalation of the electronic sector and increased consumption of electronic products [2]. Globally, a significant portion of the plastic waste generated ends up in landfills due to limited plastics management. Landfills were recognized as the primary source of plastic losses to the environment [3]. The speed with which products become obsolete further aggravates problems with the environment, which becomes the destination of waste generated and disposed of improperly in most cases. In this sense, urban prospecting techniques can serve to mitigate the negative impact that anthropic products cause on the environment [4].

Each year, approximately 50 million tons of electronic waste are generated worldwide. This electronic waste contains hazardous materials harmful to human health and the environment [5]. There is also a lack of knowledge available to control methods and strategies for handling electronic waste. Moreover, there is no agreement between the reuse, recycling, recovery, and reform industries. However, recycling is the main method for dealing with electronic waste [6].

The ISWA—International Solid Waste Association, a non-governmental and nonprofit international association based in Austria that works to promote and develop solid waste care worldwide, published a report stating that in 2019 the world broke the record for the production of electronic waste with 53.6 million tons. This number is equivalent to the production of e-waste of 7.3 kg per inhabitant in one year. In Europe, that number reached 16.2 kg per inhabitant. According to the report, Asia generated the most massive volume of e-waste in 2019—around 24.9 Mt, followed by the Americas (13.1 Mt) and Europe (12 Mt), while Africa and Oceania generated 2.9 Mt and 0.7 Mt, respectively [7].

Europe is firmly committed to recycling plastics, especially packaging [8], and according to the European waste management hierarchy, the preferred way for reuse is through recycling. From an environmental point of view, reuse is beneficial if the impacts that arise during a certain period of use of a reused product are less than the duration of a new product. Otherwise, reuse is not beneficial for recycling [9]. From the four white goods (washing machine, refrigerator, stove, and freezer) and four small electronic devices (PC—personal computer, printer, monitor, and laptop), by using Life Cycle Assessment, 68% by weight of Germany e-waste might be reused [9].

In the United States, more than 35 state manufacturer's extended liability laws on waste were adopted between 1991 and 2009. For electronics, these laws vary between early disposal fees or return mandates [10]. In Indonesia, a study aims to understand influential factors for its residents to participate in a formal e-waste recycling program. Electronic waste collection requires collaboration between the government and companies of electronics manufacturers and must be supported by a legal framework [11].

A model that correlates the amount of electronic waste with the economic increase was developed by [12]. According to the researchers, there is a strong linear correlation between the global generation of electronic waste and the Gross Domestic Product—GDP. It was observed that electronic waste be considered an opportunity for the recycling or recovery of valuable metals (for example, copper, gold, silver, and palladium), given their significant content in precious metals than in ores.

Cucchiella et al. [13] stand that an economic assessment will define the potential revenues from the recovery of 14 e-products, for instance: notebooks with LCD—liquid crystal display and LED—light-emitting diode display; TVs with CRT—cathode-ray tube, LCD, and LED; cell phones, and smartphones; photovoltaic panels; HDDs—hard disk drives; and SSDs—solid-state drives and tablets) based on current and future available volumes, which is one of the main challenges in the recycling sector.

1.2. Generation of Electronic Waste in Brazil and Motivation to Manufacture WEEE Interlocking Plastic Pavers

According to the UN—United Nations report, in 2017, Brazil generated about 1.4 Mt of WEEE's—Waste Electrical and Electronic Equipment [14]. Brazil is the second-largest producer of this waste in the Americas, with the U.S.A. being the first one, and Brazil leading the rankings in Latin America [15].

Brazil has an increasing rate of electronic waste generation. With few treatment systems for e-waste, the material goes to landfills or goes into informal chains in gray markets. It is necessary to implement a reverse logistics system within a regulatory and technical framework. There are possible synergies between the demands of formal companies and the instruments that will be applied to a recently approved regulation as a way to overcome these limitations [16]. Some solutions have been developed. To assess sustainability and prioritize alternatives for potential implementation in the metropolitan region of Rio de Janeiro, a hybrid scheme for collecting WEEE with delivery points at stores, subway stations, and neighborhood centers was designed, with the involvement of private companies, cooperatives, and social enterprises; and complete recycling of all components [17].

In Brazil, the National Solid Waste Policy was established through Law n° 12,305 of 2010. This law aims at the non-generation, reduction, reuse, recycling, and treatment of solid waste and its disposal. The law reinforces the need to conduct research that contributes to reducing environmental impacts caused by the increasing generation of this waste [18]. Thus, it is essential to characterize products from the reuse of these residues. With the current waste generation scenario from the electronics industry, space is being created for new raw materials: WEEE [19]. In this case, recycling is one of the most essential actions to reduce impacts and boost production [20]. In a parallel way, it is very important the construction of an environment propitious to the creation of recycling actions, as discussed by Vieira et al. [21]

Within the scope of the Digital Inclusion Policy, the Brazilian Federal Government and partners promote the implementation of CRC's—Computer Reconditioning Centers, responsible for the correct treatment of e-waste, promoting professional qualification for young people in vulnerable social condition, the reconditioning of computers and the environmentally proper disposal of waste [14].

According to the annual report of the ABRIN—Brazilian Association of Recycling and Innovation, 986 t of computer equipment discarded by the Brazilian federal government were processed by the CRC's in 2017 [16], 36% of which consists of CPUs, 4% of notebooks or laptops, 24% of monitors, 10% of printers, 25% of other IT—Information Technology items, and 1% of IT furniture [22]. The CRC "Programando o Futuro" of Valparaíso/GO produces interlocking plastic pavers from housings and plastic components. These pavers can be used for applications on floors, walkways, elevated walkways, and parking. Figure 1 illustrates the pavers' production flow and its application.

Figure 1. (left) From e-waste to interlocking plastic pavers: the recycling way. Containers to collect e-waste are distributed in strategic points (malls, supermarkets etc.). The Computer Reconditioning Centers (CRC) also collects electronic equipment either in public or private institutions. **(right)** On a sand bed, the pavers form a pavement with space for accommodation and drainage.

It is worthy to say that the CRC program processed about 5% of total mass-produced by the federal government [10]. We think that commercial opportunities may be opened up for this action. For each ton of material disposal, based on average market data, for approximately US $0.5 per kg, there is an approximate amount of resources in the order of US $511 thousand in revenue for the CRC, if they operated in the correct commercial disposal and within the requirements of the current legislation [14].

The reuse of the ever-increasing electronic waste as a load in the manufacture of concrete beams can improve structural strength and allow its use in constructing buildings [23]. Other researchers showed that electronic waste's addition provided a decreasing trend of structural resistance [24]. Assessing the presence of other polymers applicable in pavers' production, the use of vulcanized rubber residue from unserviceable tires with residual aggregates from the demolition of civil construction and Portland cement was investigated [25].

The insertion of PET—Polyethylene Terephthalate—in concrete alloys for pavements generated a reduction in the concrete's compression strength when comparing the metered specimens with and without the aggregate. In contrast, the water absorption of parts manufactured with PET waste showed higher values than those observed for parts without waste. There are significant environmental advantages compared to traditional parts [26].

Considering the economic importance of reusing e-waste in civil construction, we proposed in the present research the characterization of those interlocking plastic pavers manufactured by the CRC "Programando o Futuro". In Brazil's case, it is typical for the creation of products for civil construction manufactured by using mining or mineral rejects since those are the country's critical economic activities. Characterization of products is a necessary step to define the benefits of recycling [27], as producing interlocking blocks with iron ore [28].

Different from the usual composites used in civil construction, where a substance to strength it is mixed into the concrete matrix, in the present work the interlocking blocks are manufactured exclusively with e-waste and some added charge, as we describe below. This is a new type of material. Thus, this fabrication process lacks the paver's characterization. This is a crucial step to the transition between an idea to a product. Re-use or recycling must show to be economical and effective. As we show in the present paper, the interlocking blocks produced from e-waste are suitable for civil construction. The attributes of interest are defined for application on walkways and parking lots. The results obtained are compared with those of other types of materials typically used in the construction industry, following international standards.

This paper is organized as follows. The next section presents the methodology for materials characterization: International Standards have been used and the metrological procedures for the calculations of uncertainties were described. All the essays are specified, and the equipment identified to allow replicability of the results. After that, the essays' results are shown and compared with those obtained for standard pavers produced from concrete, and also under the regulatory or normalized definitions for some countries. Finally, the conclusions and scenarios for future researches are presented.

2. Materials and Methods

The interlocking plastic pavers studied in this research are produced from a mixture of polymers from computers or electronic equipment with other substances. As we have pointed out above, this manufacture is different for that of reinforcement of concrete blocks. In the transformation/recycling process, 100% of the plastic waste from the component/sub-assembly can be used. These residues, composed of ABS—Acrylonitrile Butadiene Styrene, and PS—PolyStyrene polymers, make up 70% of the main processed load. The remaining 30% of the cargo comprises other polymers or residual materials from the carcasses, such as metallic residues, sawdust, discarded toys, cigarettes, and mineral powder. This 30% load can also contribute to the reuse of non-recyclable waste.

The preferred polymers for the highest percentage of load, ABS and PS, have excellent characteristics such as low density, resistance to high temperatures, good mechanical resistance, high resistance to abrasion, dimensional stability, etc. [29].

The pavers under study have inclusions and pores, which are not expected to modify an interlocking paver's mechanical behavior, as they are small in size compared to the entire material. These inclusions come from metallic materials constituting the housings of computer components or other granular materials added as cargo during production. This residue is also desirable since the manufacture of the interlocked block must be economically viable. The non-treatment of the waste, together with the primary raw material, reduces the cost of the process; that is, even with the presence of inclusions, the interlocked block must have resistance to the compression load compatible with the values defined in the Standards for the same models of artifacts of concrete.

To fuse the polymer particles' surface, a procedure known as agglutination is used, which is a technique for compacting and agglomerating polymeric and residual particles previously ground through heating and pseudo plastification of the material [30]. After this, the mixture is brought to constant pressure in a matrix made with the parameters of the block's dimension and shape. In this process, there is a reduction in the compacted material's free surface energy, minimizing the generation of pores and conditioning the stress relief on the surface atoms.

Rigid polymers, like PS, ABS, PC—Polycarbonate, can have high values for elastic modulus, ductility, and yield strength, determining the limits of tensile strength. These mechanical characteristics change with temperature. An increase in temperature produces a reduction in the modulus of elasticity and the limit of tensile strength, with a consequent improvement in ductility. For this control, polymers' melting, and glass transition, temperatures are used as essential parameters [31] for determining the agglutination temperature.

To obtain specimens for essays, 10 interlocking pavers were chosen at random from a batch produced in CRC "Programando o Futuro". The specimens have been produced in dimensions accordingly the Standards. Essays were defined from the materials' predetermined applications: resistance to compression load, surface hardness, medium absorption, medium density, resistance to micro-abrasive wear, and study of surface topography and phase contrast by SPM—Scanning Probe Microscopy. The flowchart of the essays is depicted in Figure 2.

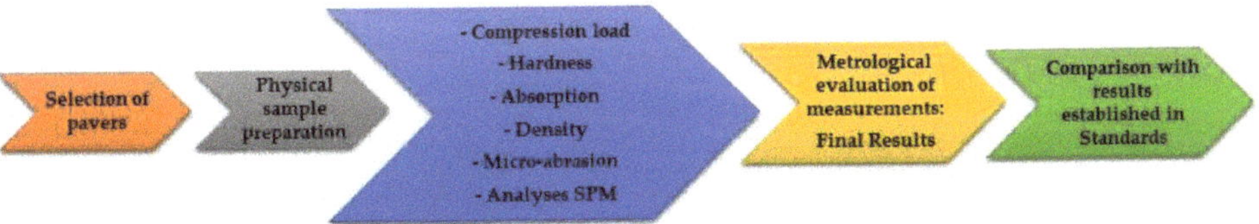

Figure 2. Process flowchart describing the steps of this work.

2.1. Manufacture of Interlocking Block

The manufacturing process of interlocking plastic pavers goes through the following steps:

- Classification: Materials containing appropriate and preferred waste for recycling are organized;
- Separation: The main raw material polymers (PS and ABS) are separated from other materials of the equipment, which might serve as cargo;
- Shredding: The separated materials go through the shredding process and are stored in specific quantities for agglutination;
- Agglutination: a compaction technique and agglomeration of previously ground particles with pseudo-plasticization [30] at temperatures between 115 °C and 140 °C. In this process, 70% PS or ABS and 30% load are used (which may be sawdust, fabric, other plastic waste, or even small metals aggregated in previously separated housings);
- Pressing: Put the agglutinated mass in a specific mold (in the form of an interlocking floor, block, or brick) and start pressing for approximately 1 min, solidifying at room temperature (which is between ~20 °C and 30 °C in Valparaíso/GO).
- Deburring: The mass is cleaned by trimming the burrs generated by the material purged in the compression.

2.2. Specimens

Since there are no specific standards for assessing the mechanical strength of interlocking floors made up of e-waste, procedures based on the following standards were adopted:

- EN ISO 604: 2002-Plastics—Determination of compression properties:
- ASTM D695-02a-Standard Test Method for compression Properties of Rigid Plastics;
- ISO 2039-2: 1987 Plastics—Determination of hardness—Part 2: Rockwell hardness
- ASTM D785-08: 2015-Standard Test Method for Rockwell Hardness of Plastics and Electrical Insulating Materials

The shape of a specimen and the method used, and the number of tests involved per essay were based on these Standards. The Standards EN ISO 604: 2002 and ASTM D695-02a are equivalent. The number of specimens to be tested depends on the materials' structure. For isotropic materials, at least five specimens from each material sample must be tested. For anisotropic materials, at least ten specimens of each sample should be tested in the perpendicular direction and five samples in the direction parallel to the main axis of the anisotropy [32].

The material produced from the process described above can be considered isotropic because there is no preferred direction in this process. Therefore, 5 interlocking blocks from the batch of 10 were randomly chosen. These were identified by blocks 3, 4, 5, 7, and 8. The specimens made to the mechanical evaluations were prepared by mechanical machining: cutting with a band saw, roughing by milling, and finishing by sanding. Figure 3 shows the steps to obtain the specimens.

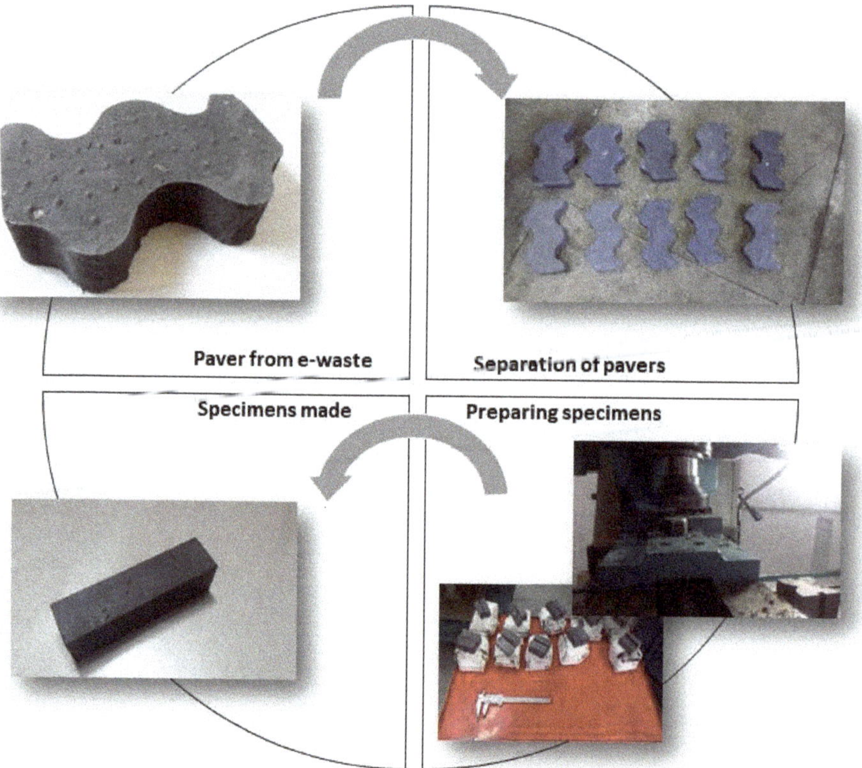

Figure 3. Separation of pavers and specimen preparation for essays.

2.3. Mechanical Assessments

2.3.1. Evaluation of Resistance to a Compression Load

The tests to evaluate resistance to compression load were carried out at the Fracture Mechanics Laboratory of the Metallurgy Department—Escola de Minas—Universidade Federal de Ouro Preto (DEMET/EM/UFOP), with a servohydraulic testing system from the manufacturer MTS, model 810 Material Test System, Software Station Manager—M15 Flex test 40—2013, 10×10^3 kg capacity with a servo valve.

To better understand the behavior of the blocks in different environmental conditions common in our country, four additional tests were carried out, classified from "a" to "d", as described below. Five specimens were prepared for each test specification.

"a": Test at room temperature (which was about 24 °C);
"b": Saturation test (24 h of immersion);
"c": Hot test (70 °C);
"d": Freezing test (0 °C).

In all the essays, the specimen is compressed parallel to its main axis at a constant speed until it breaks, or the tension decreases or when the length reaches a predetermined value [24,30].

Uncertainty Calculation

For all tests, uncertainty calculations were made accordingly to the GUM—Guide to the Expression of Uncertainty in Measurement [33]. The uncertainty calculations considered repeatability in five different specimens since there is no replication of the measurement at the same point because it is a destructive test. The main uncertainty factors are the instrument, the resolution, and the repeatability.

The sources of quantifiable uncertainty are presented in the Ishikawa diagram, as shown in Figure 4. The other possible components of uncertainty are considered non-quantifiable or do not influence the measure.

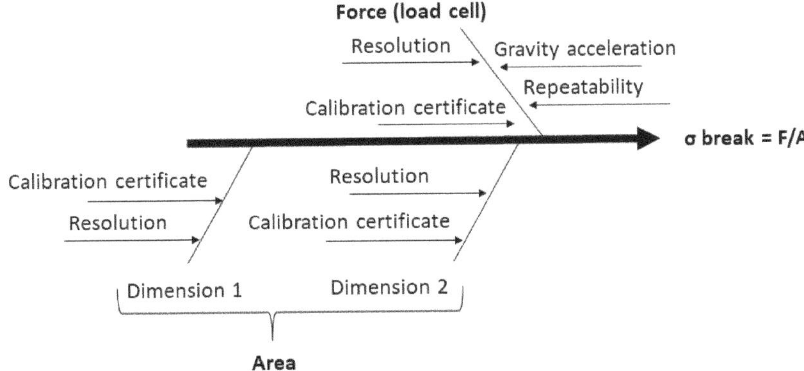

Figure 4. Cause and effect diagram for calculating measurement uncertainty in the compression strength test.

Given the above definitions, the uncertainty components' calculation is done in two steps: the repetition of measurements or Type A assessment of standard uncertainty; and uncertainty inherited from the instrument or Type B assessment of standard uncertainty.

As we conclude from the essays, the higher contribution to the uncertainty comes from the repetition of compression strength measurements. This uncertainty contribution will always exist because the material is composed of at least two different polymeric molecules and contains impurities and/or inclusions, which can cause a variation in the behavior in the different conditions. There is a broad interval in the measurements to evaluate hot, cold, and saturated conditions. Worthy to say that it is important that the confidence intervals remain under the conditions of block application.

It is important to mention that the same type of approach for calculating the uncertainties was adopted in all the test procedures described below. For space savings, they will not be described here, but are available upon request.

2.3.2. Surface Hardness Assessment

The surface hardness assessment was obtained from a branded pen impact type durometer Time, model TH130, resolution 0.1 HRC. The instrument was provided by the Machining Company Torneamentos Mariana LTDA, Mariana-MG. In this evaluation, only one specimen was used, chosen randomly, repeating 5 measurements at three different points.

2.3.3. Evaluation of Average Water Absorption

To evaluate the average absorption, first, we measure the mass of a dry sample. After this measurement, the specimen is immersed in distilled water for 24 h, and the mass measurement is repeated. The absorption evaluation was calculated in 5 samples with five mass checks each, dry and saturated. These specimens were used in the resistance test to the saturated compression load (test "b" described above). The absorption value is obtained in percentage by the expression given by the ratio between the difference between wet and dry mass [34].

The measurements used an analytical balance: Shimadzu—Model AY220, resolution 0.0001 g, provided by the Laboratory of Polymers and Electronic Properties of Materials LAPPEM/ICEB/UFOP.

The block absorption analysis, since one does not have a proper standard, used the procedures for interlocking concrete blocks. This technique evaluates the increase of mass in a solid specimen with a porous appearance, suggesting water accumulation in the pores, which explains the variation in mass and, consequently, the absorption [35].

In the evaluation of average density, the ratio between the mass and the corresponding volume of the test pieces tested in the evaluation of average absorption was used. The volume was calculated from the Mitutoyo 300 mm universal caliper measurements, resolution 0.05 mm, provided by the Machining Company Torneamentos Mariana LTDA, Mariana-MG.

2.3.4. Evaluation of Resistance to Micro-Abrasive Wear

In the evaluation of resistance to micro-abrasive wear, two specimens were used. In each specimen, 5 tests were carried out; that is, there are 10 wear evaluation points. The tests were performed on equipment developed by the Foundry Laboratory DEMET/EM/UFOP. A Scanning Electron Microscope was used to measure the wear cap—MEV, from the manufacturer TESCAN, model VEGA 3 SEM, of the Laboratory NANOLab EM/UFOP.

Wear was assessed by rotating a sphere against a static specimen, with abrasive mud between the sphere and the specimen. The spherical cap-shaped groove has a volume calculated from the sphere's radius and the spherical cap's diameter. Thus, the wear coefficient can be interpreted as the volume of material worn per unit of force and per unit of sliding distance (m^3/Nm) [36].

After micro-abrasive essays, a study of the wear mechanism was carried out by SEM—Scanning Electron Microscope.

2.3.5. Study of Surface Topography and Phase Contrast with SPM

For the analysis of the microform of the material structure and the combination of the compound's different materials, an analysis of the surface topography was done using topography and phase-contrast images. An AFM—Atomic Force Microscope—was used, from the manufacturer Park Systems, model Park XE7, from the Microscopy Center of Instituto Federal de Minas Gerais—IFMG, Ouro Preto.

The mode used in this test was tapping (intermittent contact), because a soft material with energy dissipation favors the distinction between attractive and repulsive regimes for image generation.

The set of techniques allows generating images, where the channels monitored in the tapping mode were topography, error, and NCM Phase in trace and retraction (a round trip of the AFM probe in different images). In the form of peaks and valleys, dimensional information on the surface morphology and phase difference of materials are observed [37].

3. Results and Discussions

The techniques for evaluating resistance to compression stress, surface hardness, absorption rate, density, and resistance to micro-abrasive wear were used in order to expose blocks produced from WEEE to conditions or application requests: as an interlocking block for pavements, walkways, parking, and walkways.

3.1. Resistance to Compression Load

A fragile fracture was observed in all samples. The compression load resistance values have been considered satisfactory when compared to those obtained for interlocking concrete blocks. The samples that presented lesser resistance showed a fracture in portions with some inclusion. As we have discussed above, the material does not treat the separation of aggregate residues. The material under study had a maximum breaking strength limit of approximately 9355 N.

The stress versus strain curves showed features of essays on polymers, with a linear relationship between stress/strain in some interval, which allows the evaluation of Young's module. The value is within the expected parameters for polymers, even for specimens treated at high temperatures, low temperatures, and saturation. As we have described above, five specimens have been tested for each essay. Figure 5 shows the stress-strain results for each essay. Results for saturated specimens, as well those for high temperatures, showed significant dispersion. The essay at room temperature showed lower dispersion, as shown in the uncertainty value presented in Table 1.

Table 1. Values of average compression strengths and average Young modules.

Compression strength at room temperature	(42 ± 5) MPa
Saturated compression strength	(46 ± 14) MPa
High temperature compression strength	(27 ± 12) MPa
Frozen compression strength	(44 ± 12) MPa
Young's modulus at room temperature	(1.420 ± 0.013) GPa
Saturated Young's modulus	(0.950 ± 0.008) GPa
Hot Young's modulus	(1.110 ± 0.006) GPa
Frozen Young's modulus	(1.150 ± 0.004) GPa

The average values were considered in the uncertainty calculations for each measurement, as discussed above. The values of average compression strengths and average Young modules with respective estimated uncertainties are expressed in Table 1.

There is a reduction in resistance to compression in the case of the hotter specimen, and a small variation in Young's modulus concerning the other tests. This occurs due to the relative movement of adjoining polymer chains as tension increases, facilitated by reducing secondary bonding forces on a molecular scale. The material becomes softer and more ductile [31].

Comparing the results of the polymeric paver's compression load resistance with those required for concrete pavers, as shown in Table 2, the material has acceptable behavior.

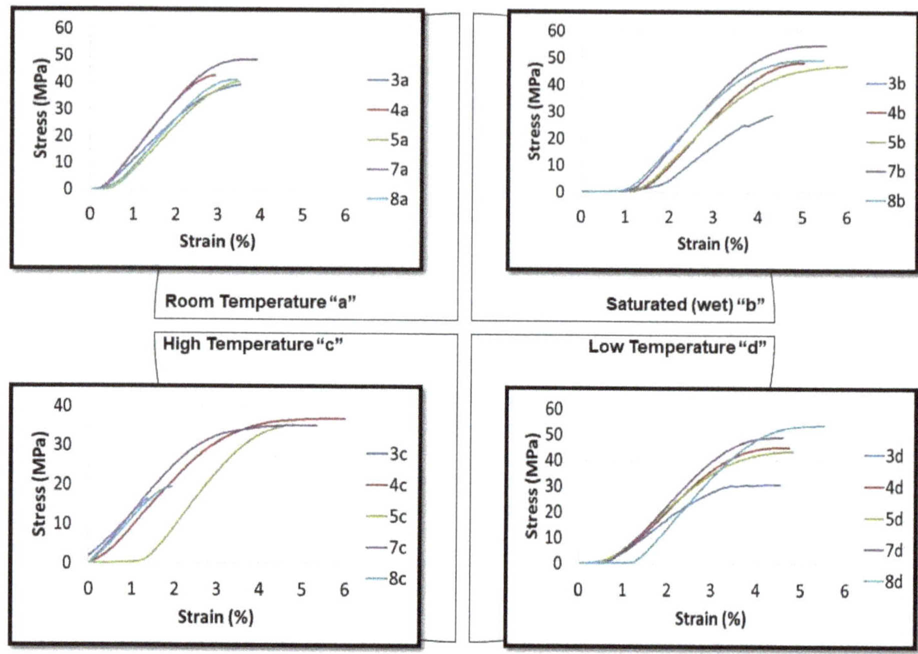

Figure 5. Stress x Strain graphs. The results for each test are presented for the four specimens obtained from each block (numbered as 3, 4, 5, 7, and 8, randomly selected as described above). Letters "a", "b", "c", and "d" refers to the different essays.

Table 2. Comparative table of resistance to compression load.

Standard		Requirements				
		Light Traffic and Pedestrians	**Heavy Traffic**	**Bike Paths and Parking**	**Special Vehicles**	**Acceptable Limits**
ABNT NBR 9781:2013	Brazil	$\geq$35 MPa	$\geq$50 MPa	-	-	-
ASTM C936:1996	USA	-	-	-	-	$\geq$55 MPa
CSA A231.2-95	Canada	-	-	-	-	$\geq$50 MPa
SANS 1058:2009	South Africa	$\geq$25 MPa	$\geq$50 MPa	-	-	-
AS/NZS 4456.4:2003	Australia	$\geq$25 MPa	$\geq$35 MPa	$\geq$15 MPa	$\geq$60 MPa	-
BSEM 1388:2003	Europe	No individual results <3.6 MPa and breaking load <250 N/mm				

Compression strength of the Interlocking Block made from WEEE

Room temperature (MPa) = 37–47
Saturated (MPa) = 32–60
Hot (MPa) = 15–39
Frozen (MPa) = 32–56

3.2. Surface Hardness

The material under study has high surface hardness because it has ABS in its composition. The evaluation took place at room temperature (~24 °C), and the values are shown in Table 3 below. Each column refers to each essay's three points: at the left, middle, and right parts of the specimen. The three columns at left present the measurements obtained directly, and the three columns at right present the results after corrections, which have been calculated from the calibration certificate. As one can observe, the several values obtained for a specific point are not the same, due to micro deformations on the tested surface, even though it is a non-destructive evaluation.

Table 3. Result table of surface hardness values.

Point Measurement (Left, Middle, Right)			Point Correction (Left, Middle, Right)		
45.6	52.7	49.1	46.9	54.0	50.4
48.6	52.6	42.7	49.9	53.9	44.0
50.5	52.2	46.8	51.8	53.5	48.1
51.9	47.9	50.5	53.2	49.2	51.8
51.7	52.6	50.9	53.0	53.9	52.2
Point-corrected average hardness (HRC)			51.0	52.9	49.3
Average hardness (HRC)			51.1		

The average surface hardness measurement with respective estimated uncertainty can be expressed as = (51.1 ± 1.0) HRC.

Since the hardness assessments in Portland concrete are currently made by axial compression with a sclerometer obtaining the sclerometric index (MPa), it is not possible to directly compare the results obtained in HRC for this work. Winslow [38] used the ASTM Standard E 18-79 "Rockwell Hardness and Rockwell Superficial Hardness of Metallic Materials", obtaining the concrete's average surface hardness ranging from 45 to 52 HR. Thus, the polymer block surface hardness meets Portland concrete specifications applied on floors, since we have obtained values between 50.1 and 52.1 HRC

3.3. Average Water Absorption

Construction with interlocking blocks requests the existence of space between them, in the resulting application. This gap is necessary to allow rainwater to flow to the soil and be absorbed; the parking lot should not be impermeable. Thus, a necessary feature of pavers is the low absorption of water, allowing it to flow to the soil. For the interlocking blocks produced from e-waste, a low absorption rate was obtained, as shown in Table 4. This low rate indicates conditions for the flow of rainwater, and the construction of parking with these blocks, giving space between them for free movement, settlement, and drainage of water to the soil passing through the gap.

Table 4. Comparative table of absorption rate.

Standard		Requirements
ABNT NBR 9781:2013	Brazil	≤6%
ASTM C936:1996	USA	Average: ≤5%
BSEM 1388:2003	Europe	<6%
Interlocking Block Made from WEEE		A (%) = 0.081–0.089

The tests showed that the material has properties consistent with light paving, such as parking lots, sidewalks, bike paths, and walkways. The average value of the absorption rate was considered in the measurement of uncertainty calculations. With the respective estimated uncertainties, we have = (0.085 ± 0.004)%.

3.4. Average Density

The material under study has a low average density compared to the density of concrete that varies between 2.5 and 3 g/cm^3, as shown in Table 5.

Table 5. Result table of average density values.

Specimen	3b	4b	5b	7b	8b
	8.2142	9.6173	8.3386	9.5825	8.4978
	8.2142	9.6173	8.3386	9.5825	8.4978
Measures (g)	8.2142	9.6173	8.3386	9.5825	8.4978
	8.2142	9.6173	8.3386	9.5825	8.4978
	8.2142	9.6173	8.3386	9.5825	8.4978
Individual average (g)	8.2142	9.6173	8.3386	9.5825	8.4978
Volume (cm^3)	7.99	8.18	8.11	8.83	8.01
Density (g/cm^3)	1.02806	1.17571	1.02819	1.08522	1.06090
Average density (g/cm^3)			1.075615		

The average density measurement with respective estimated uncertainty can be expressed as = (1.075615 ± 0.000019) g/cm^3.

3.5. Resistance to Micro-Abrasive Wear

A circumferentially shaped cavity was generated, shown in a $35\times$ zoom in Figure 6A. An inclusion is observed in this cavity, as o observed from the $130\times$ zoom, shown in Figure 6B. The wear mechanism is uniform, even in the presence of different materials in the compost. The details on the wear surface and confirming the wear mechanism by scratching are shown in Figure 6C, where micropores can be observed. The scratch wear mechanism was observed, generating micropores. For this test, Aluminum Carbide was used as an abrasive, suitable for evaluating severe wear, resulting in uniformity as detected in Figure 6D. We assume that these micropores allow the penetration of contaminants, which might accelerate the material's degradation.

In the micro-abrasion resistance assessment equipment, the measurement replication was done in the same specimen, but at different points. The analyzes of the resistance to micro-abrasive wear were done in 2 specimens conditioned to loads of 1 N and 3 N, respectively. The evaluation of resistance to micro-abrasive wear showed values adequate to the designed application. Table 6 shows the loss of volume from the dimensions of the wear cavities.

Table 6. Comparative table of resistance to micro-abrasion.

Standard		Requirement
ABNT NBR 12042:1992	Brazil	Group A: floor with high traffic demand $\leq$0.8 mm Group B: heavy pedestrian traffic between 0.8 mm and 1.6 mm Group C: light traffic between 1.6 mm and 2.4 mm
ASTM C 936-1996	USA	Volume loss: $\leq$15 cm^3/50 cm^2
BSEM 1388:2003	Europe	<23 mm
Interlocking block made from WEEE		K with load 1 N (m^3/Nm) = 0.0008 to 0.0012 K with load 3 N (m^3/Nm) = 0 to 0.0004

The final results of the average micro-abrasive wear coefficients for the 1 N and 3 N loads with the respective standard uncertainties are:

- K with load 1 N = (0.0010 ± 0.0002) m^3/Nm.
- K with load 3 N = (0.0002 ± 0.0002) m^3/Nm.

ABNT, ASTM, and BSEM standards are based on wear depth, even by different tests. Thus, to correlate the methods used, we calculate the respective depths of the spherical caps formed by the wear sphere:

- K with load 1 N: 0.48 mm
- K with load 3 N: 0.35 mm

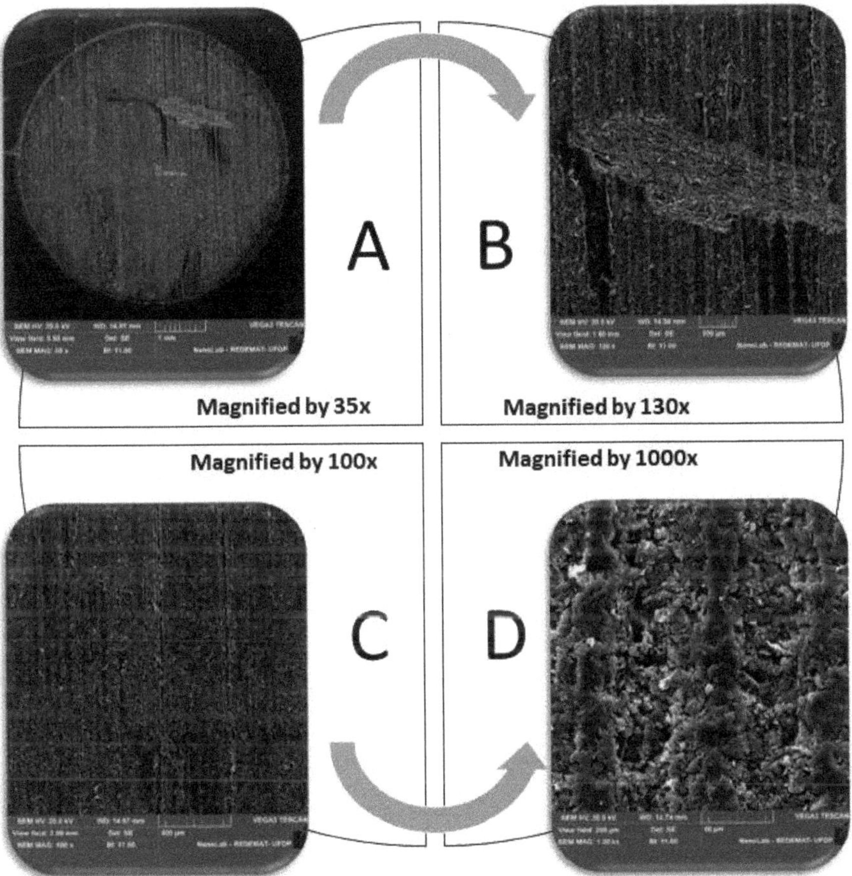

Figure 6. Magnified by (**A**) 35×, (**C**) 100×, (**B**) 130×, and (**D**) 1000×. Evaluation of micro-abrasive wear in inclusions using the SEM technique.

By using the same methodology as this project, [36] obtained a wear coefficient K between 9.67×10^{-13} m^3/Nm and 19.02×10^{-13} m^3/Nm in the evaluation of abrasive wear resistance of grinding bodies produced with white cast iron high chrome. It is a metallic material with an average hardness of 80 HRC. Thus, we assume that the polymer under study has resistance to micro-abrasion equivalent to that expected.

3.6. Surface Topographic Analysis and Phase Contrast with SPM—Scanning Probe Microscopy

To analyze the microform of the material structure and the connection between the compound's different materials, an analysis of the surface topography was done using topography and phase-contrast images. Using the AFM—Atomic Force Microscopes technique, with the Force-Tip Sample ratio, topography, error, and phase-contrast images are generated in trace and retraction (round trip of the probe in different images), where it is observed in shape of peaks and valleys dimensional information of surface morphology and phase difference of materials [37].

The mode used in this test was Tapping (Intermittent contact), because the soft material with energy dissipation favors the distinction between attractive and repulsive regimes for image generation. According to [39], AFM can distinguish the different modules of elasticity of the constituent materials even though there is no difference in the sample's topography. This is thanks to the AFM needle that vibrates on the sample and detects the material's ability to absorb the energy of the shock with the needle. As the sample was manufactured from materials already processed, this mode will bring morphological and topographic information after reproducing the specimens' polymers.

It is possible to observe variations in the surface due to "voids", that is, variations in the measurements of the "high" and "low" points. The lower the darker surface (bottom of the "hole"): in this case, black. The higher, the clearer.

The red ruler in Figure 7, highlighted in Figure 7A, represents a depth of approximately 510 nm. The green ruler, represented in the graph of Figure 7A, has a point that stands out for its light color and in the lower left corner a dark color, showing a "very high" point and a "very low" point in approximately 560 nm.

As the probe identifies different materials by the attractive/repulsive action, detecting the material's ability to absorb the energy from the shock of the needle, Figure 8 shows the presence of more than one material in the compound under study, with blue being the largest quantity, the red represents the base for fixing the sample. A reasonable degree of homogeneity of the polymeric matrix under study is observed.

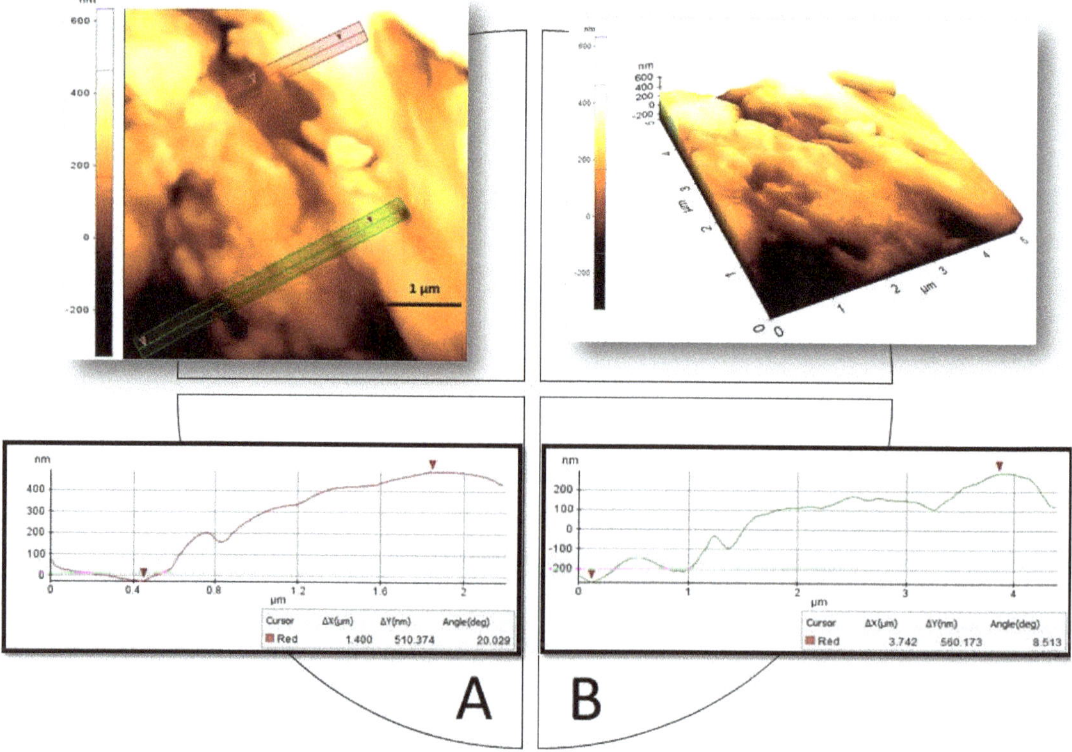

Figure 7. Topographic images of the Atomic Force Microscope (AFM) in trace of the probe. At left viewpoint vertical to the image and, at right, and oblique view. The two plots below show topography for two regions: (**A**) refers to the red scale of the image, and (**B**) refers to the green scale of the image.

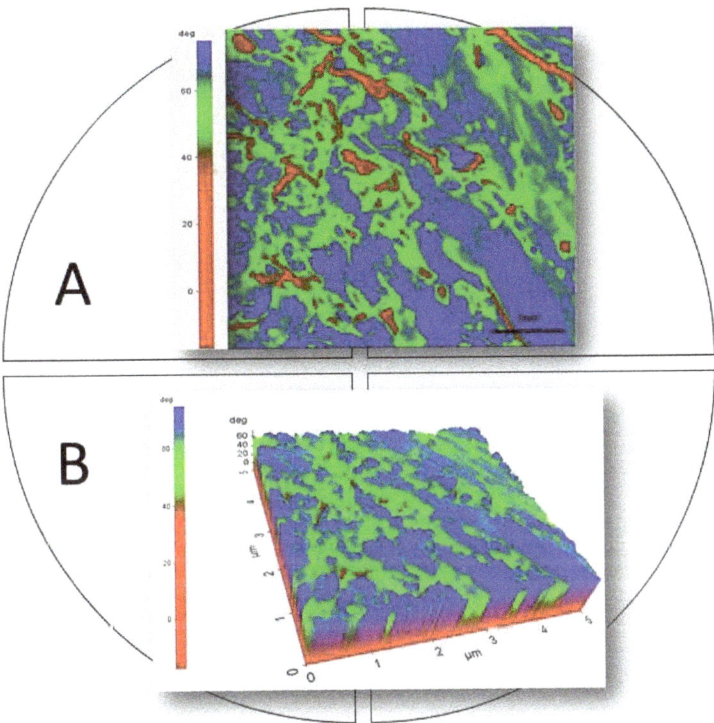

Figure 8. (**A**) Phase contrast images of the sample in 2d. (**B**) Phase contrast images of the sample in 3d.

4. Conclusions

Production of electronic waste increases rapidly, and it is necessary the definition of public policies to deal with this. Recycling is one important line of action and recycling industrial residues for the production of basic materials for civil construction is a growing way to use these residuals. To this production be effective and economical, it is necessary for the characterization of those new materials.

In this work, we characterize the mechanical properties of interlocking plastic pavers produced with residues from the electric-electronic industry. These pavers are produced at the Computer Reconditioning Center Programando o Futuro, in the city of Valparaíso/GO. CRC is part of the public policy of the Ministry of Communications, which aims to the recondition equipment, the adequate disposal of e-waste, and education and training of young people in situations of social vulnerability.

The CRC developed a technique to produce pavers from e-waste. As we have discussed in the introduction, this type of action is adherent to other researches that aim to manufacture products for civil construction from the electrical and electronic industry residuals.

Pavers are produced from a mixture of different materials, but their main components—in the order of 70%—are computer housings composed of ABS, PS, and PC. The extra charge of material crushed with cigarettes, CDs, toys, or makeup packs is added. This mixture is heated at temperatures between 115° and 140° and pressed into molds.

Characterization of materials for civil constructions obtained from mineral or steel industries residuals, or even with the re-use of those originally used in the civil construction, is common in the literature. However, since the recycling of polymers aiming at civil construction is very recent, one needs a description of the basic features of those new materials obtained from e-waste. This is also necessary to support public policies, as well as to define new Standards for these new materials.

Tests of resistance to compression, hardness, absorption, and micro-abrasion were performed. The topographic characteristics of the material were also verified, allowing us to visualize the structural arrangement of its components. From the tests carried out and the estimated uncertainty assessments, it is clear that interlocking plastic pavers manufactured from e-waste behave similar to blocks made from concrete. These evaluations showed that the material has properties consistent with those required for light paving, such as parking lots, sidewalks, bike paths, walkways, and walkways.

Future works and researches should be developed in the definition of Standards for this new type of material, as well as in the development of new methods for their characterization. We believe that this kind of work, mainly in Brazil or other developing countries, which produce tons of non-reused electronic equipment and where recycling of e-waste is a newborn activity, should influence public and private agents to support the creation of new cheap products for diverse applications. It is also necessary to create new essays, describing physical, chemical, and biological features.

This type of recycling or reconditioning action contributes to at least two of the Sustainable Development Goals proposed by the UN [40] which must be implemented by all countries in the world by 2030.

Author Contributions: L.T.G.: conceptualization, testing and essays, data curation, investigation, methodology, and writing original draft. R.F.B.: supervision, investigation, writing—review and editing, and methodology. A.T.B.: conceptualization, investigation, formal analysis, methodology, writing—review and editing, and project administration. All authors have read and agreed to the published version of the manuscript.

Funding: This research was carried out with partial funding of the Coordination for the Improvement of Higher Education Personnel—Brazil (CAPES): Financing code 001; Funding from the Brazilian agencies CNPq and Fapemig, and financial support from PROPP-UFOP are also acknowledged.

Data Availability Statement: Publicly available datasets were analyzed in this study. This data can be found at: http://www.repositorio.ufop.br/handle/123456789/12341 (accessed on 20 January 2021).

Acknowledgments: We acknowledge support from Leonardo Godefriod, Maria Aparecida Pinto and Geraldo Lúcio de Faria for helping us in some essays and granting access to their laboratories. Further, we acknowledge the support from Geraldo Alves (Torneamentos Mariana) and Danúbia Clemente (SENAI Mariana) for giving access to their facilities. Paulo Couto and Alexandre Mendes are also acknowledged for discussions and clarifying issues on the uncertainties assessments. Finally, we are very grateful to the people of CRC Programando o Futuro, in the person of its coordinator Vilmar Simion for preparing the samples and giving us access to the data of their production.

Conflicts of Interest: The authors declare no conflict of interest.

References

1. Geyer, R.; Jambeck, J.R.; Law, K.L. Production, Use, and Fate of All Plastics Ever Made. *Sci. Adv. (Calif.)* **2017**, *3*, e1700782. Available online: http://advances.sciencemag.org/content/3/7/e1700782 (accessed on 5 February 2019). [CrossRef] [PubMed]
2. Chowdhury, A.; Patel, J. E-Waste Management and its Consequences: A Literature Review. *Prestig. E-J. Manag. Res.* **2017**, *4*, 52–63.
3. Yadav, V.; Sherly, M.A.; Ranjan, P.; Tinoco, R.O.; Boldrin, A.; Damgaard, A.; Laurent, A. Framework for Quantifying Environmental Losses of Plastics from Landfills. *Resour. Conserv. Recycl.* **2020**, *161*, 104914. [CrossRef]
4. Corrêa, H.L.; Figueiredo, M.A.G. Urban Mining: A Brief Review on Prospecting Technologies. *Braz. J. Dev.* **2020**, *6*, 29441–29452. [CrossRef]
5. Khutale, Y.P.; Yadav, S.B.; Awati, R.V.; Awati, V.B. Electronic Waste Management. *Int. J. Recent Trends Eng. Res.* **2019**, *5*, 42–50. [CrossRef]
6. Mao, S.; Gu, W.; Bai, J.; Dong, B.; Huang, Q.; Zhao, J.; Zhuang, X.; Zhang, C.; Yuan, W.; Wang, J. Migration of heavy metal in electronic waste plastics during simulated recycling on a laboratory scale. *Chemosphere* **2020**, *245*, 125645. [CrossRef]
7. ISWA. The Global E-Waste Monitor 2020. Available online: http://www.iswa.org/home/news/news-detail/article/-21c8325490/109/ (accessed on 15 July 2020).
8. Demets, R.; Roosen, M.; Vandermeersch, L.; Ragaert, K.; Walgraeve, C.; Meester, S. Development and Application of an Analytical Method to Quantify Odour Removal in Plastic Waste Recycling Processes. *Resour. Conserv. Recycl.* **2020**, *161*, 104907. [CrossRef]

9. Boldoczki, S.; Thorenz, A.; Tuma, A. The Environmental Impacts of Preparation for reuse: A Case Study of Weee Reuse in Germany. *J. Clean. Prod.* **2020**, *252*, 119736. [CrossRef]
10. O'reilly, P.G. Regulation, Product Durability, and Market Process in Recycling Electronic Waste. Ph.D. Thesis, Mineral And Energy Economics at the Colorado School of Mines, Faculty and the Board of Trustees of the Colorado School of Mines, Golden, CO, USA, 2019. Available online: http://mountainscholar.org/bitstream/handle/11124/174005/OReilly_mines_0052E_11872.pdf?sequence=1&isAllowed=y (accessed on 27 July 2020).
11. Siringo, R.; Herdiansyah, H.; Kusumastuti, R.D. Underlying Factors behind the Low Participation Rate in Electronic Waste Recycling. *Global J. Environ. Sci. Manag.* **2020**, *6*, 203–214. Available online: http://www.gjesm.net/article_37686_ee1060947420a30acb5e857c61c2a433.pdf (accessed on 27 July 2020).
12. Awasthi, A.K.; Cucchiella, F.; D'Adamo, I.; Li, J.; Rosa, P.; Terzi, S.; Wei, G.; Zeng, X. Modelling the Correlations of E-Waste Quantity with Economic Increase. *Sci. Total Environ.* **2018**, *613–614*, 46–53. [CrossRef] [PubMed]
13. Cucchiella, F.; D'Adamo, I.; Lenny Koh, S.C.; Rosa, P. Recycling of WEEEs: An economic assessment of present and future e-waste streams. *Renew. Sustain. Energy Rev.* **2015**, *51*, 263–272. [CrossRef]
14. ABRIN. Brazilian Association of Recycling and Innovation-Annual Report 2017-Disposal Data, Management, Indicators and Collection. São Paulo: 65p. 2018. Available online: http://www.abrin.eco.br/ (accessed on 13 May 2019).
15. Baldé, C.P.; Forti, V.; Gray, V.; Kuehr, R.; Stegmann, P.; The Global E-Waste Monitor. United Nations University (UNU), International Telecommunication Union (ITU) & International Solid Waste Association (ISWA), Bonn/Geneva/Vienna: 2017. Available online: http://collections.unu.edu/view/UNU:6341 (accessed on 13 May 2019).
16. Santos, K.L. The Recycling of E-Waste in the Industrialised Global South: The Case of Sao Paulo Macrometropolis. *Int. J. Urban Sustain. Dev.* **2020**, 1–14. [CrossRef]
17. Souza, R.G.; Clímaco, J.C.N.; Sant'anna, A.P.; Rocha, T.B.; Valle, R.A.B.; Quelhas, O.L.G. Sustainability assessment and prioritisation of e-waste management options in Brazil. *Waste Manag.* **2016**, *57*, 46–56. [CrossRef]
18. MMA–Ministry of the Environment. National Solid Waste Policy. Available online: http://www.mma.gov.br/cidades-sustentaveis/residuos-solidos/politica-nacional-de-residuos-solidos (accessed on 28 July 2020).
19. Xavier, L.H.; Resíduos Eletroeletrônicos na Região Metropolitana do Recife (RMR): Guia Prático para um Ambiente Sustentável. 1ª edição. *Recife: Editora Massangana.* 2014. Available online: https://www.researchgate.net/publication/271587261_Residuos_Eletroeletronicos_na_Regiao_Metropolitana_do_Recife_Guia_pratico_para_um_ambiente_sustentavel (accessed on 13 May 2019).
20. Hopewell, J.; Dvorak, R.; Kosior, E. Plastics recycling: Challenges and Opportunities. Philosophical Transactions of The Royal Society B: Biological Sciences. *R. Soc.* **2009**, *364*, 2115–2126. [CrossRef] [PubMed]
21. De Oliveira Vieira, B.; Guarnieri, P.; Camara e Silva, L.; Alfinito, S. Prioritizing Barriers to Be Solved to the Implementation of Reverse Logistics of E-Waste in Brazil under a Multicriteria Decision Aid Approach. *Sustainability* **2020**, *12*, 4337. [CrossRef]
22. MCTIC: Ministry of Science, Technology, Innovations and Communications. Computers for Inclusion. 2019. Available online: http://www.mctic.gov.br/mctic/opencms/indicadores/detalhe/O-Centro-de-Recondicionamento-de-Computadores-2.html (accessed on 10 March 2019).
23. Manjunath, B.T.A. Partial Replacement of E-plastic Waste as Coarse-Aggregate in Concrete. Procedia Environ. *Sci.* **2016**, *35*, 731–739. [CrossRef]
24. Hamsavathi, K.; Prakash, K.S.; Kavimani, V. Green High Strength Concrete Containing Recycled Cathode Ray Tube Panel Plastics (E-Waste) as Coarse Aggregate in Concrete Beams for Structural Applications. *J. Build. Eng.* **2020**, *30*, 101192. [CrossRef]
25. Silva, A.P. *Fabricação de Análise de Pavimentos Intertravados (PAVERS) Utilizando Resíduos de Borracha de Pneus Inservíveis e Resíduos de Construção Civil e Demolição (RCD) Como Agregado Miúdo;* 29 f. TCC (University Graduate)-Curso de Engenharia Civil; Universidade do Vale do Paraíba: São José dos Campos, Brazil, 2016.
26. Paschoalin Filho, J.A.; Pires, G.W.M.O.; Rezende, L.V.S.; Santana, J.C.C. Resistência a Compressão e Absorção de água de peças de piso Intertravado Manufaturadas com Resíduos de pet. *HOLOS* **2019**, *1*, 1–21. [CrossRef]
27. Jain, N.; Garg, M.; Minocha, A.K. Green Concrete from Sustainable Recycled Coarse Aggregates: Mechanical and Durability Properties. *J. Waste Manag.* **2015**, 1–8. [CrossRef]
28. Silva, F.L.; Moreira, G.F.; Coelho, H.C.S.; Pires, K.S.; Kruger, F.L.; Araujo, F.G.S. Caracterização Quantitativa Eletrônica de Blocos Intertravados com Rejeito de Minério de Ferro. XXVIII Encontro Nacional de Tratamento de Minérios e Metalurgia Extrativa. Belo Horizonte-MG. 2019. Available online: https://www.artigos.entmme.org/?wpfb_file_sort=%3Cfile_size (accessed on 7 January 2020).
29. Dias, M.R.Q. Separação de Plásticos de REEE Recorrendo à Separação Gravítica e Flutuação por Espumas. IST Técnico Lisboa. Engenharia Geológica e de Minas. Lisboa. 2013. Available online: http://fenix.tecnico.ulisboa.pt/downloadFile/395146045208/Disserta%C3%A7%C3%A3o_MaraDias_MEGM_54230.pdf (accessed on 28 May 2020).
30. Wassermann, A.I. Processamento e Características Mecânicas de resíduos Plásticos Misturados. Universidade Federal do Rio Grande do Sul–Departamento de Engenharia Química. Porto Alegre. 2006. Available online: http://lume.ufrgs.br/bitstream/handle/10183/17360/000714892.pdf;jsessionid=4BE30884E5DF1BECC0AC870E76A497E4?sequence=1 (accessed on 31 June 2019).
31. Callister, W.D.J.; Rethwisch, D.G. *Materials Science and Engineering: An Introduction*, 8th ed.; John Wiley and Sons: Hoboken, NJ, USA, 2009; Available online: https://www.amazon.com/Materials-Science-Engineering-Introduction-8th/dp/0470419970 (accessed on 20 January 2021).

32. ISO-International Standardization Organization. *604:2002: Plastics-Determination of Compression Properties*; ISO: Geneva, Switzerland, 2012.
33. BIPM: Bureau International des Poids et Mesures. *JCGM 100:2008 GUM, 1995 with Minor Corrections Evaluation of Measurement Data—Guide to the Expression of Uncertainty in Measurement*; Paris, France, 2010. Available online: https://www.bipm.org/en/publications/guides/gum.html (accessed on 20 January 2021).
34. ABNT-Brazilian Association of Technical Standards. *ABNT NBR 9781/2013: Concrete Paving Units-Specification and Test Methods*; ABNT/MB 3483: Rio de Janeiro, Spain, 2013.
35. Kassman, A.; Jacobson, S.; Erickson, L.; Hedenqvist, P.; Olsson, M. A new test method for the intrinsic abrasion resistance of thin coatings. *Surf. Coat. Technol.* **1991**, *50*, 75–84. [CrossRef]
36. Santos, L.C. *Influência do Tratamento Térmico de Revenimento Sobre a Resistência ao Desgaste Abrasivo de Corpos Moedores Produzidos Com Ferro Fundido Branco de Alto Cromo*; 59 f. TCC (University graduate)-Engenharia Metalúrgica; Universidade Federal de Ouro Preto: Ouro Preto, Spain, 2014.
37. Leite, F.L.; Zeimath, E.C.; Herrmann, P.S.P. Análise de Minerais do Solo por Espectroscopia de Força Atômica. *EmbrapaSão Carlos-Sp* **2005**, *70*, 1–7. Available online: http://www.infoteca.cnptia.embrapa.br/bitstream/doc/28204/1/CT702005.pdf (accessed on 26 October 2018).
38. Winslow, D.N. A Rockwell Hardness Test for Portland Cement Concrete. Purdue University-West Lafayette, Indiana. 1981. Available online: http://pdfs.semanticscholar.org/c4f4/a5d61aa8017b453ce85533c011ccbec8262b.pdf?_ga=2.251600705.615054613.1599741451-287955651.1599741451 (accessed on 10 September 2020).
39. Herrmann, P.S.P.; da Silva, M.A.; Bernardes, F.R.; Job, A.E.; Colnago, L.A.; Frommer, J.E.; Mattoso, L.H. Microscopia de Varredura por Força: Uma Ferramenta Poderosa no Estudo de Polímeros. Polímeros: Ciência e Tecnologia, São Paulo. 1997, Volume 1, pp. 51–61. Available online: http://www.scielo.br/pdf/po/v7n4/8878.pdf (accessed on 13 May 2019).
40. UN. The 2030 Agenda for Sustainable Development. 2015. Available online: http://sdgs.un.org/goals (accessed on 23 January 2020).

Article

Assessment of Eco-Friendly Pavement Construction and Maintenance Using Multi-Recycled RAP Mixtures

David Vandewalle [1,2], Vítor Antunes [3,4,*], José Neves [3] and Ana Cristina Freire [4]

1   Department of Civil Engineering, Ghent University, St. Pieternieuwstraat 33, 9000 Ghent, Belgium; djvdwall.vandewalle@ugent.be

2   Department of Civil Engineering, Architecture and Georesources, Instituto Superior Técnico, Universidade de Lisboa, Av. Rovisco Pais, 1049-001 Lisboa, Portugal

3   Department of Civil Engineering, Architecture and Georesources, CERIS, Instituto Superior Técnico, Universidade de Lisboa, Av. Rovisco Pais, 1049-001 Lisboa, Portugal; jose.manuel.neves@tecnico.ulisboa.pt

4   LNEC, National Laboratory of Civil Engineering, Av. do Brasil 101, 1700-066 Lisboa, Portugal; acfreire@lnec.pt

*   Correspondence: vitorfsantunes@tecnico.ulisboa.pt

Received: 16 July 2020; Accepted: 12 August 2020; Published: 14 August 2020

Abstract: The demand for more sustainable solutions has led an ever-growing number of stakeholders to being committed to pursue the principles of sustainability in pavement management. Different stakeholders have been looking for tools and methodologies to evaluate the environmental impacts of the solutions, for which the life cycle assessment (LCA) proved to be an appropriate methodology. This paper is focused on the LCA of road pavement multi-recycling based on the use of bituminous mixtures with high rates of reclaimed asphalt pavement (RAP). In order to promote the circular economy, a comparative analysis was performed on a road pavement section by taking into account different scenarios, which stem from the combination of production, construction and rehabilitation activities incorporating different RAP rates in new bituminous mixtures: 0% (as reference), 25%, 50%, 75% and 100%, respectively. LCA results have been expressed in terms of four damage categories: human health, ecosystem quality, climate change and resources. Results have shown that both recycled and multi-recycled bituminous mixtures lead to substantial benefits in comparison with the solution employing virgin materials, hence embodying a sustainable approach. The benefits grow with the increase in the RAP rate with an average decrease of 19%, 23%, 31% and 33% in all the impact categories for a 25%, 50%, 75% and 100% of RAP rate.

Keywords: LCA; road pavement management; RAP; multi-recycling; circular economy; sustainability

1. Introduction

The road infrastructure is one of the most important and omnipresent assets in construction engineering. In fact, roads are not only an important part of a society's transportation network, but also a public asset in overall terms. Indeed, the transport of goods and people worldwide is mainly done by road infrastructures.

Consequently, in view of the importance of these infrastructures and the need to guarantee appropriate both short and long-term behaviors of the road networks worldwide, several tonnes of bituminous mixtures are produced every year. In 2007, 1.6 trillion tonnes of bituminous mixtures were produced worldwide [1]. Around the world, pavement construction companies are required to provide end-products complying with both the high standards, defined by road authorities, and the sustainability criteria. The life cycle assessment (LCA) is a methodology intended to assess the environmental impacts associated with all stages of the life cycle of a solution, or product, during the manufacture, distribution, usage and after recycling or final disposal in a landfill, similarly to the operations associated with road pavement construction.

Bituminous courses have a limited usage period, due to ageing and other deterioration phenomena, and must hence be replaced after this period. Therefore, every year huge budgets are allocated to road pavement management (construction, maintenance and rehabilitation actions). During maintenance and rehabilitation actions of bituminous pavements, huge quantities of reclaimed asphalt pavement (RAP) are produced. This secondary raw material can be recycled in similar applications. RAP is defined as the removed asphalt materials containing bitumen and aggregates, which result from reconstruction and resurfacing activities, or from actions intended to obtain access to buried utilities or also from reject and surplus production [2].

Due to fairly recent developments, such as the increase in energy costs, as well as accrued environmental concerns and common sustainability awareness, there has been a shift in focus when it comes to evaluating assets that are procured and used. Governments are trying to minimize both the life cycle costs and the environmental impact, instead of simply considering the lowest cost alone when it comes to achieving added value from constructed assets [3,4].

A rising number of different stakeholders are taking account of sustainability issues in their management and business activities, namely the ones associated to the road industry. This allows stakeholders to implement in their decision-making process other key aspects that, until now, had not been seen as fundamental, such as the life cycle economic, environmental and social impacts. For this the multi-criteria decision analysis had been used. The recent guidelines issued for 2050 the construction sector presupposed by those for 2020. They continue to pursue and increase the sustainable competitiveness of the sector and deal with further challenges [5].

Pavement asset management actors have been pursuing further sustainable solutions and practices to adopt in their construction and maintenance actions [6–8]. The general approach for improving sustainability consists of reducing energy consumed, emissions generated and the amounts of virgin material used. This means implementing preventive maintenance, lowering the bituminous mixture heat and adopting other eco-friendly pavement technologies.

One of the most common methods to address these concerns and correctly quantify emissions, as well as the consumption of materials and energy, is the LCA [9–14].

This paper aims to present an attributional LCA of alternative solutions incorporating high RAP recycling rates in new bituminous mixtures. Considering the limited life period of the bituminous solutions, more than one-time RAP recycling was taken into consideration. Therefore, this paper focuses on: (1) multi-recycling solutions with high RAP incorporation rates to provide a clear understanding on how RAP and the multi-recycling of flexible pavements affect the pavement's life-cycle; (2) modelling of the recycled mixture's production phase considering the RAP's processing necessary to have greater control over the mixture's properties and performance, and consequently increase the incorporation rates. For the latter, it was assessed the impacts of each process directly related with the production of the bituminous mixture. Both analyses were performed by assessing four impact damage categories: human health, resource consumption, climate change and ecosystem quality; together with 15 impact factors. To perform such an analysis, a real road section with real implemented solutions was compared with the alternative solutions.

2. Background

LCA allows investigating the environmental aspects of products, services or any activity associated with them, by categorizing and measuring the contributions during the flow, since the beginning up to the final product, from a life cycle perception. The LCA method should preferably be totally inclusive, insofar as it should assess the full environmental impact and include all processes from a cradle-to-grave approach. For a pavement LCA, this means that it begins with the extraction of raw materials and ends with their landfill disposal, when a conventional linear economy is considered.

The LCA methodology was standardized in the ISO 14040 series that divides the LCA framework into four steps [15,16]: Goal and scope definition (ISO 14040: 1997); Inventory assessment (ISO 14040: 1997); Impact assessment (ISO 14042: 2000) and Interpretation (ISO 14043: 2000).

Although the above-mentioned standards were combined in EN ISO 14040:2006 [17], the general LCA framework described in this standard still integrates the same four steps. Actually, the methodology proposed in this study follows this general framework.

When innovative materials and/or solutions are under development, the use of LCA tools allows assessing the gains or losses in terms of environmental impacts. These tools are appropriate to measure the benefits, in terms of reduction in the use of raw materials and non-renewable resources, when recycled materials are included in the adopted solutions [15,18–27].

The European Union issued some strategies to encourage the recycling and reuse of by-products and wastes. The Directive 2008/98/EC established that all EU member states should define guidelines to support the re-use of products and to prepare for re-use activities, with a view to achieve a high-quality recycling. RAP recycling in similar applications, as a secondary raw material, promotes the material circular economy that is included in the European Commission package issued in December 2015 [28]. It establishes that, in a circular economy, the value of the products is perpetuated for the maximum length of time. When a product achieves its end-of-life, it will be re-used to create further value, i.e., a cradle-to-cradle application, reducing the landfill needs and the consumption of natural resources [3,7,29,30].

The ever-growing procurement for greener and more sustainable solutions has been the basis for the adoption of RAP recycling solutions in bituminous mixtures. Bituminous mixture recycling is a progress toward achieving sustainable pavement systems [10,31–39]. The paving recycled solutions provide an important contribution to the paving industry to attain a sustainable economic and environmental development. This diminishes the depletion of quality resources in landfills (e.g., bitumen and aggregates), which are finite and costly.

RAP recycling in similar applications has environmental and economic benefits [32,40]. The savings in the amount of necessary aggregates and binder and the consequent reduction in their transportation to the plant site leads to economic benefits. The savings in raw materials have a direct impact on the reduction in fuel emission and consumption, during the operations of extraction, processing and transportation of these materials. Furthermore, it leads to a decreased need for non-renewable resources, landfill transportation and space for the disposal of used pavement materials [7,41,42].

Since the first application of RAP in new bituminous mixtures, several studies have been developed using different RAP incorporation rates, mixing techniques and recycling agents or rejuvenators [34,37,38,43–62]. Different RAP recycling rates, varying between 0 and 100%, were found in the literature for different application courses, e.g., base to surface courses. Several real applications, using in-plant and in-place recycling techniques, which presented good results, have been reported since the 1970s. It must be noted that, in these years, the knowledge about recycling techniques and appropriate plants was still underdeveloped. However, even in these conditions, similar or greater performances were obtained when recycled and traditional courses were compared [54]. Generally, those studies revealed that the in-service performance of recycled mixtures was in line with the laboratory results. No RAP treatment was applied in the cases where the recycled mixtures exhibited a worse performance than the traditional mixes with virgin materials [37,63–65]. With proper methodology and plant upgrades, the high RAP incorporation solutions eventually led to the current application to pavement construction and rehabilitation.

3. Methodology

3.1. General Procedure

This paper evaluates high RAP recycling rates in bituminous mixtures and its multi-recycling by applying an LCA to a real road section. Moreover, the RAP processing to allow higher incorporation percentages was modelled in the production phase to assess the additional impacts of the activities and the remaining activities (Figure 1). The focus, in this case, lies on 15 impact factors that can be broken down into four damage categories: human health (especially for construction workers and people directly involved in the production of flexible pavements), climate change, ecosystem quality and use of resources.

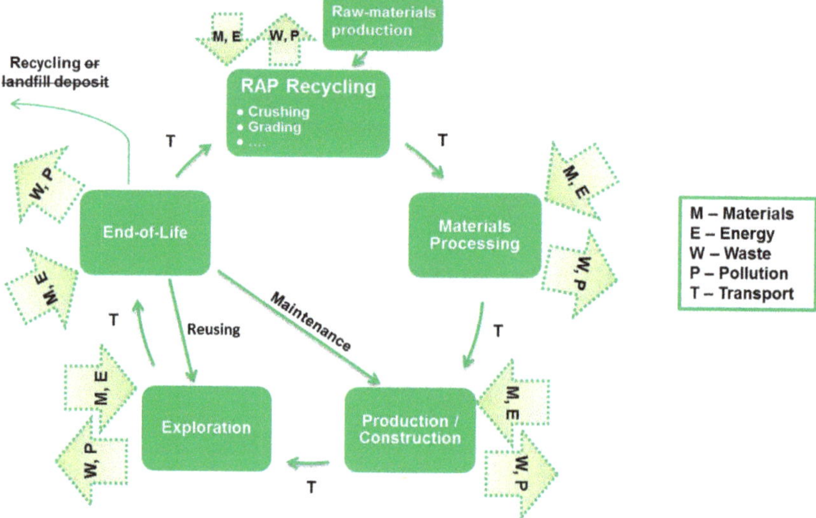

Figure 1. Phases and components of pavement life cycle assessment (LCA).

The LCA presented in this study was performed using the SimaPro software from PRé Sustainability in combination with the Ecoinvent database. This software-database combination is one of the most widely disseminated LCA tools, and provides the user with an interface, several comprehensive environmental information databases and various methods to perform the impact assessment [20]. The software provides a clear interface to perform a full LCA, comprising the four steps described in EN ISO 14040: 2006 [17]. The Ecoinvent database is the most widely used in the construction sector. The database used was the Ecoinvent 3—allocation, cut-off by classification—unit base. The underlying philosophy of the cut-off approach is that the producer does not obtain any credit for the provision of recyclable materials, but is nonetheless fully responsible for their disposal as waste [66].

3.2. Functional Unit

The LCA functional unit creates the basis for comparing the different structure scenarios with the same utility for an equivalent function [12,67,68]. In this case, it involved a pavement unit that carried the same number of vehicles per year, over the same project analysis period (PAP). This period comprised the different periods referred to by different authors and specifications [6,9,29,67,69–73]. The functional unit presented in this LCA was a real 1 km-long road section of a two-lane roadway, one in each direction, with an individual width of 3.5 m each and a total PAP of 69 years, from 1946 to 2015. The geometric characteristics, as well as the different mixtures used for each course, are presented in Figure 2. This figure also displays the overall maintenance and rehabilitation (M&R) strategy applied in reality by roadway concession holder (tasks, courses and application scheduling) and the 2 main structure scenarios considered in this LCA.

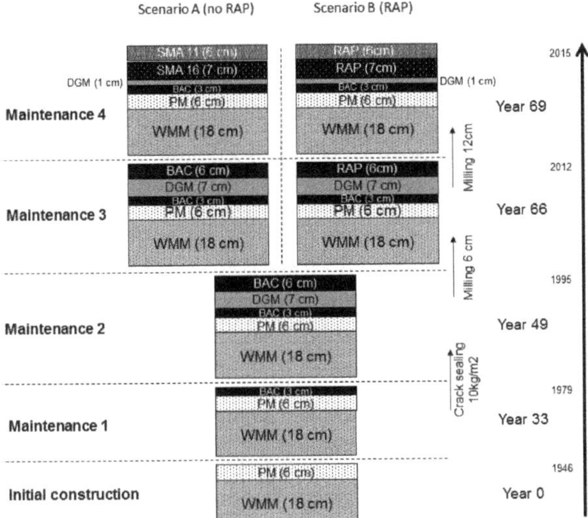

Figure 2. Flexible pavement structure and maintenance and rehabilitation strategies. WMM: wet mix macadam (hydraulic material); PM: penetration macadam; DGM: Dense-graded bituminous mixture; BAC: Bituminous asphalt concrete; SMA 16: stone mastic asphalt 16; Binder course PMB 45/80-65; SMA 11: stone mastic asphalt 11; Surface course PMB 45/80-65; RAP: Reclaimed asphalt pavement.

The first three phases of the pavement construction and maintenance history: initial construction, maintenance 1 and maintenance 2, were equal for both scenarios (A and B). Equal construction solutions and materials were also considered. Thus, these construction and maintenance actions were not taken into account in the study. For the remaining maintenance tasks, equal dates for the maintenance actions were assumed for both scenarios. However, these scenarios presented different solutions for the bituminous mixture to be applied. As such, the scenario A represents the real road section with the applied solutions and maintenance action performed by the roadway concession holder; whereas the scenario B is a proposed alternative scenario to be evaluated in the study. Scenario B considered alternative solutions using hot recycled bituminous mixtures assessed assuming the same geometry, traffic increasing and maintenance periods applied in real road section (scenario A), as presented in Figure 2. The maintenances 3 and 4, applied in scenario A and simulated in scenario B, were as follows:

- Maintenance 3 implied the milling of the bituminous dense-graded mixture (AC) of the surface course and its replacement by a new surface course. In scenario A, this course consisted of a virgin AC, while in scenario B this course incorporated virgin AC, together with a certain rate of RAP that was obtained from the milling using a rejuvenator percentage.
- Maintenance 4 implied the milling of both upper courses and their replacement by a new binder and surface course, respectively. In the case of scenario A, these courses consisted of stone mastic asphalt (SMA) mixtures with virgin materials, while in the courses in scenario B a certain RAP rate was incorporated together with virgin AC using a rejuvenator percentage.

To assess the use of RAP material in new bituminous mixtures, different incorporation rates were considered in this LCA. Scenario B was broken down into 4 sub-scenarios, hereafter referred to as "LCSx", as can be seen in Table 1. Scenario A was subsequently renamed LCS0 to simplify comparison in the remaining course of the LCA.

Table 1. Definition of studied scenarios.

Scenario	RAP (%) *	Reference
Scenario A	0	LCS0
Scenario B	25	LCS1
	50	LCS2
	75	LCS3
	100	LCS4

* mass percentage.

3.3. System Description and Boundaries

Both base and surface courses were set as the boundaries of the LCA analysis. Due to the insufficient and outdated data regarding wet mix macadam (WMM) and penetration macadam (PM) courses, these were not considered in the LCA. Given the comparative nature of this LCA and the fact that these two courses were present throughout all scenarios, this omission had no relative impact on the LCA outcome. The following boundaries were subsequently included:

- The construction of the courses, as well as the rehabilitation and maintenance activities, were limited by the previously defined boundaries and were in conformity with the dates and course thicknesses of the real section (Scenario A).
- Raw materials extraction needs to produce the mixtures applied in those courses.

All the transportation needs as refers to the transport of raw materials from the suppliers to the bituminous plant, and from the bituminous plant to the work-site and vice-versa. Considering the RAP system boundaries, a distinction was established between the pre- and post-processing of the RAP. A 'cut-off' allocation approach, as described in [6], implied that only the post-processing of recycled material, such as RAP should be accounted for by the system. In this case, this would imply that any environmental impact, resulting from the milling and hauling of RAP, would be excluded from the system. These processes, however, were already an inherent part of the considered M&R activities of the pavement and were, therefore, accounted for in the LCA.

3.4. Collected Data

Two types of data are essential for an LCA [74]:

- Primary data, which is specific for the production, processes of the product or service. This data is obtained from the goods' producers, and the operators of processes and services, as well as their associations.
- Secondary data presents generic and/or average data for the studied solution, considering the products and operations. Secondary data can be obtained from the sources of primary data, sometimes with some modifications, and from national databases, consultants and research groups.

For this study, the data was selected to be as representative of the Portuguese conditions during the PAP as possible. The sources for these data included national road authorities, active construction companies and technical experts. The data was mostly related to the inventory analysis of raw material extraction and production, fuels, construction and transportation vehicles and machines. This information was obtained primarily from the Ecoinvent database, but was modified whenever possible and appropriate to approach the Portuguese reality. The different data were combined in SimaPro to model the life cycles for the five different solutions defined as LCSx, where x is the number of the solution [75,76].

4. Life Cycle Inventory

4.1. General Implementation

The life cycle inventory (LCI) stage consisted of the actual data collection and modelling of the system. First, this paragraph briefly discusses both the functionality and the modelling features of SimaPro. Secondly, it addresses the data sources, the calculations that were performed to provide significant input and the modelling of the distinct phases of each LCSx in SimaPro.

4.2. Extraction Phase

The extraction phase of materials comprised the two sub-phases as follows: production and transport of virgin materials. To calculate the required amount of virgin aggregate and virgin binder for each course, several calculations were performed. The starting point of these calculations was the primary data, obtained from a variety of sources, in combination with the characteristics of the functional unit, the length and width of the unit of pavement and the depth of each course.

First, to determine these amounts, a fixed binder-aggregate ratio of 95-5 was determined for all mixtures, except for the SMA mixtures, which have a 93.5-6.5 ratio. This implies that, regardless of the technical specifications, the majority of mixtures contains 95% aggregate and 5% binder. Furthermore, to derive the amount of virgin aggregate and binder from the total amount of material, the RAP incorporation percentage for each specific course was determined. Table 2 shows the Ecoinvent unit processes associated with the production of virgin aggregate and binder, as well as the transport of these materials by truck.

Table 2. Material extraction processes and corresponding SimaPro unit processes.

Process Definition	SimaPro Unit Process
Production of virgin aggregate	Gravel, crushed (RoW) | production
Production of virgin bituminous binder	Pitch (RoW) | petroleum refinery operation
Transport of aggregate and binder	Transport, freight, lorry > 32 metric tonne, EURO4 (GLO)

The second sub-phase of the extraction of materials involved the transport of virgin material to the bituminous mixing plant. The selected unit process for transport by truck in SimaPro, "Transport, freight, lorry > 32 metric tonnes, EURO4 (GLO), has a default unit of tonne-kilometre (tkm)". To define this amount for each course, the total mass of transported material and the distance between the production site and the mixing plant were calculated. A distance of 100 km and 150 km was considered from the quarry to the mixing plant and from the binder supplier to the mixing plant, respectively.

4.3. Production Phase

The mixture production phase addresses the environmental impacts related to the production of the different mixtures considered in the system. It was assumed that all the mixtures were produced at a conventional heavy fuel oil (HFO) fired batch mix plant. To account for variations in composition, mixing temperature, moisture content of aggregates and initial temperature of raw materials, of the several types of mixtures, the thermal energy (TE), required to produce the different bituminous mixtures, was determined according to Equation (1) [67].

$$TE = \left[\begin{array}{c} \sum\limits_{i=0}^{M} m_i \times C_i \times (t_{mix} - t_0) + m_{bit} \times C_{bit} \times (t_{mix} - t_0) + \sum\limits_{i=0}^{M} m_i \times W_i \times C_w \times (100 - t_0) \\ + L_v \times \sum\limits_{i=0}^{M} m_i \times W_i + \sum\limits_{i=0}^{M} m_i \times W_i \times C_{vap} \times (t_{mix} - 100) \end{array} \right] \times (1 + CL) \quad (1)$$

where,

- TE is the thermal energy (MJ/tonmixture) necessary to produce one ton of bituminous mixture;

- m_i is the mass of aggregates of fraction i;
- C_i is the specific heat capacity coefficient of aggregate fraction i;
- M is the total number of aggregate fractions;
- t_{mix} is the mixing temperature of a bituminous mixture;
- t_0 is the ambient temperature;
- m_{bit} is the mass of bitumen;
- C_{bit} is the specific heat capacity coefficient of bitumen;
- W_i is the water content of aggregates of fraction i;
- C_w is the specific heat capacity coefficient of water;
- L_v is the latent heat required to evaporate water;
- C_{vap} is the specific heat capacity coefficient of water vapour;
- CL is the casing losses factor.

The casing losses (CL) is defined as the thermal energy that is lost by heating plant iron, instead of being used to heat the mixture components [21]. This factor was considered the same for all mixtures presented in this study, in accordance with the findings presented by Santos et al. (2018).

The different parameter values were based on literature and on average values in real practice [21,67,77].

Table 3 shows the values of the parameters used to calculate the thermal energy. For the mixtures, the standardized values were applied according to EN 12697-35 [77].

Table 3. Parameter values of thermal energy.

Parameter	Definition	Value	Unit
t_0	Ambient temperature	15	°C
C_{agg}	Specific heat of virgin aggregates	0.74	kJ/kg/°C
W_{agg}	Water content of aggregates	3	%/m_{agg}
C_{RAP}	Specific heat of RAP	0.74	kJ/kg/°C
C_w	Specific heat of water at 15 °C	4.1855	kJ/kg/°C
L_v	Latent heat of vaporization of water	2256	kJ/kg
C_{vap}	Specific heat of water vapour	1.83	kJ/kg
C_{bit}	Specific heat of bitumen	2.093	kJ/kg/°C
CL	General casing losses factor	27	%

The actual environmental impact, resulting from the production of the bituminous mixtures, was modelled by the SimaPro unit process "heat production, heavy fuel oil, at industrial furnace 1 MW | heat, district or industrial, other than natural gas | cut-off, U" from the Ecoinvent database.

In addition to the environmental damage, resulting from the heating of the mixtures, the mixture production phase of courses incorporating RAP mixtures, also partly accounted for the impact of RAP processing. The RAP processing sub-phase was divided into pre-processing and post-processing. The pre-processing of RAP was therefore attributed to the construction and M&R phases, while the post-processing was attributed to the mixture production phase. The post-processing of RAP consisted of four activities (crushing, stacking, conveying and screening), and was considered to have a combined capacity of 184 tonnes per hour [6].

To account for the environmental impact of RAP production in SimaPro, the four main RAP production activities, mentioned above (crushing, stacking, conveying and screening), were modelled in SimaPro by different unit processes, as Table 4 shows. As this unit process is generally expressed in hours (hr), the total amount of hours required for producing each course was thus used as input for both activities in SimaPro.

Table 4. Reclaimed asphalt pavement (RAP) processing SimaPro unit processes.

Process Definition	SimaPro Unit Process
Crushing of RAP by a crushing unit	Rock crushing {RER}| processing	Cut-off, U
Stacking of RAP by a wheel loader	Excavation, skid-steer loader {RER}| processing	Cut-off, U
Conveying of RAP on a conveyor belt	Machine operation, diesel, ≥74.57 kW, high load factor {GLO}
Screening by a mobile screener	Machine operation, diesel, ≥74.57 kW, high load factor {GLO}

4.4. Construction and M&R Phases

The construction and M&R phases have considerable environmental impacts on the system that need to be assessed. These impacts namely result from the machine operations involved in the construction of the courses and from the transportation of bituminous mixtures and milled material. The data concerning the activities and machine efficiencies were based on the literature review and on consultations to contractors and road experts.

Following the methodology used by other authors [67], machine operations were modelled in SimaPro using two different unit processes from the database, depending on the power of the machinery used. To describe the light machine operations, the SimaPro unit process 'Machine operation, diesel, ≥18.64 kW and <74.57 kW, high load factor' was used. The heavy machine operations were modelled by the unit process 'Machine operation, diesel, ≥74.57 kW, high load factor'. For the transportation by truck, the 'Transport, freight, lorry > 32 metric tonnes, EURO4 | cut-off, U' was adopted.

4.5. Work-Zone Traffic Management and Use Phases

The work-zone traffic management phase accounts for the differential fuel cost and emissions released by on-road vehicles, due to congestions generated during M&R actions, in comparison to those generated during normal road operation [67]. Considering the nature of the analysis and the fact of the study evaluating a real section of road, due to the lack of realistic data, as refers to traffic management during maintenance phases, the environmental impacts of this phase were not included.

The use phase takes account of the environmental impacts, which result from the interaction of the pavement with the vehicles, on the environment and humans throughout its PAP [67]. Other factors have been considered during the usage phase of the pavement, namely: pavement-vehicle interaction, traffic flow, leachate, carbonization and lighting [67].

In this study, however, the usage phase was not taken into account, although the contribution of this phase to the overall environmental impact of the pavement life cycle is potentially significant (or even dominant) [78]. Again, the reason for this omission lies on the lack of well-documented information regarding the case-study. However, such omission did not have significant impacts, due to the comparative basis of the analysis performed.

4.6. End-of-Life Phase

The end-of-life (EOL) phase of a pavement includes the destination of the pavement after its PAP. There are two main possible destinations for a given pavement: (1) remaining in place, or, (2) removal [67]. In this study, the pavement was assumed to remain in place and to undergo maintenance 4. All environmental impacts of this phase were therefore considered in the extraction of materials, in the mixture production, in construction, and in M&R phase.

5. Results and Discussion

5.1. Life Cycle Impact Assessment

The third step of the LCA, the life cycle impact assessment (LCIA), addresses the relationship between the inventory and the exterior; what kind of effect the system has on humans and the environment. It consisted of the following steps: classification, characterization and optional normalization, grouping, and weighting [17]. The midpoint impact category and endpoint normalization results were calculated by applying the life cycle impact method IMPACT 2002+. IMPACT 2002+ is a combination of four methods: IMPACT 2002 [22], Eco-indicator 99 [79], CML [80] and IPCC [23].

5.2. LCA Global Results

To evaluate the high RAP incorporation rates, and more specifically the multi-recycling rehabilitation, five rehabilitation scenarios for the functional unit were considered, which were referred to as LCS0, LCS1, LCS2, LCS3 and LCS4 (Table 5). Classification is the first step of the LCIA, where each elementary flow is assigned to a certain impact category, in consonance with the substances' potential for this category. In the second step (characterization), the individual emissions from the elementary flows that contribute to a single impact category are summed up. Firstly, however, the emissions are converted into indicators, using factors calculated by the IMPACT 2002+ model, in order to account for their relative contribution to a certain impact category. Table 5 shows the total results for each scenario.

Table 5. Characterization results for each scenario.

Impact Factor	Unit	LCS0	LCS1	LCS2	LCS3	LCS4
Carcinogens	kg C_2H_3Cl eq	4.93×10^3	3.99×10^3	3.66×10^3	3.33×10^3	3.13×10^3
Non-carcinogens	kg C_2H_3Cl eq	6.61×10^3	5.39×10^3	4.83×10^3	4.26×10^3	3.89×10^3
Respiratory inorganics	kg PM2.5 eq	5.68×10^2	4.65×10^2	4.24×10^2	3.82×10^2	3.57×10^2
Ionizing radiation	Bq C^{14} eq	1.13×10^7	8.46×10^6	7.61×10^6	6.76×10^6	6.47×10^6
Ozone layer depletion	kg CFC^{11} eq	2.90×10^{-1}	2.15×10^{-1}	1.93×10^{-1}	1.72×10^{-1}	1.66×10^{-1}
Respiratory organics	kg C_2H_4 eq	3.34×10^2	2.56×10^2	2.30×10^2	2.04×10^2	1.94×10^2
Aquatic ecotoxicity	kg TEG water	6.37×10^7	4.85×10^7	4.35×10^7	3.87×10^7	3.65×10^7
Terrestrial ecotoxicity	kg TEG soil	2.05×10^7	1.65×10^7	1.49×10^7	1.32×10^7	1.21×10^7
Terrestrial acid/nutri	kg SO_2 eq	1.03×10^4	8.43×10^3	7.74×10^3	7.02×10^3	6.58×10^3
Land occupation	m^2org.arable	1.02×10^4	8.87×10^3	7.81×10^3	6.74×10^3	5.80×10^3
Aquatic acidification	kg SO_2 eq	3.50×10^3	2.82×10^3	2.61×10^3	2.40×10^3	2.29×10^3
Aquatic eutrophication	kg PPO_4 P_{lim}	1.35×10^2	1.02×10^2	9.12×10^1	8.07×10^1	7.64×10^1
Global warming	kg CO_2 eq	4.58×10^5	3.84×10^5	3.58×10^5	3.32×10^5	3.15×10^5
Non-renewable energy	MJ primary	2.48×10^7	1.83×10^7	1.65×10^7	1.47×10^7	1.41×10^7
Mineral extraction	MJ surplus	6.21×10^3	5.24×10^3	4.59×10^3	3.93×10^3	3.36×10^3

Figure 3 shows the relative decrease in the characterization indicator for each scenario compared to the baseline scenario LCS0. The total average decrease for LCS1, LCS2, LCS3 and LCS4, was 20%, 28%, 36% and 40%, respectively. As can be concluded from this figure, the use of RAP and multi-recycling led to a considerable decrease in the environmental impact across all characterization factors. Moreover, this decrease was proportionate to the rate of RAP considered in the scenario; all indicators decreased as the rate of RAP increased. For the LCS4 scenario, the biggest decrease can be observed in the mineral extraction category (46%), while the global warming category displays the smallest decrease (31%).

The second part of the LCIA consisted of a damage assessment. According to ISO standards, this step is optional for an LCIA and is strongly similar to the characterization one. In the damage assessment, however, the categories were defined from an endpoint approach, in contrast to the midpoint approach used in the characterization step. Each damage category was thus compiled from several impact categories. The damage categories provided by the IMPACT 2002+ method, together with their units, are presented in Table 6. This table also includes the different impact categories assigned to each damage category.

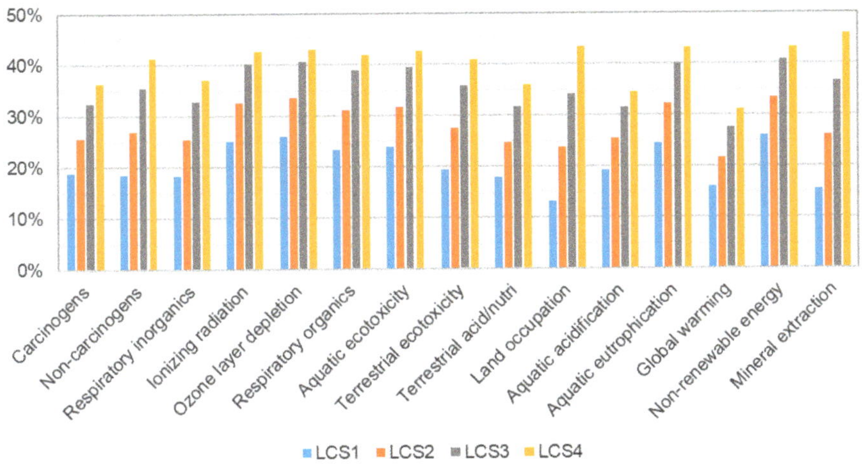

Figure 3. Characterization factors.

Table 6. Damage category definition [81].

Impact Category	Damage Category	Unit	Unit Definition
Carcinogens Non-carcinogens Respiratory inorganics Respiratory organics Ionizing radiation Ozone layer depletion	Human health	DALY	Disability-adjusted life years: characterizes the disease severity, accounting for both mortality (years of life lost due to premature death) and morbidity (the time of life with lower quality due to an illness, e.g., at the hospital)
Aquatic ecotoxicity Terrestrial ecotoxicity Terrestrial acid/nutri Land occupation Aquatic acidification Aquatic eutrophication	Ecosystem quality	$PDF \times m_2 \times yr$	The Potentially Disappeared Fraction of species over a certain amount of m^2 during a certain amount of year.
Global warming	Climate change	$kg\ CO_2$ eq	$kg\ CO_2$ equivalents into air
Non-renewable energy Mineral extraction	Resources	MJ primary	MJ primary non-renewable

Table 7 shows the total damage assessment results for each scenario. As can be expected from the characterization results, a similar decrease in impact can be observed in all damage categories for scenarios comprising RAP. Furthermore, as Figure 4 shows, this decrease is again proportional to the percentage of RAP included in the mixtures. The biggest damage category decrease was observed for resources (43% for LCS4), and the smallest for climate change (31% for LCS4).

Table 7. Total damage assessment results.

Damage Category	Unit	LCS0	LCS1	LCS2	LCS3	LCS4
Human Health	DALY	0.434	0.354	0.323	0.291	0.271
Ecosystem quality	$PDF \times m^2 \times yr$	1.88×10^5	1.52×10^5	1.36×10^5	1.21×10^5	1.11×10^5
Climate change	$kg\ CO_2$ eq	4.58×10^5	3.84×10^5	3.58×10^5	3.32×10^5	3.15×10^5
Resources	MJ primary	2.48×10^7	1.84×10^7	1.65×10^7	1.47×10^7	1.41×10^7

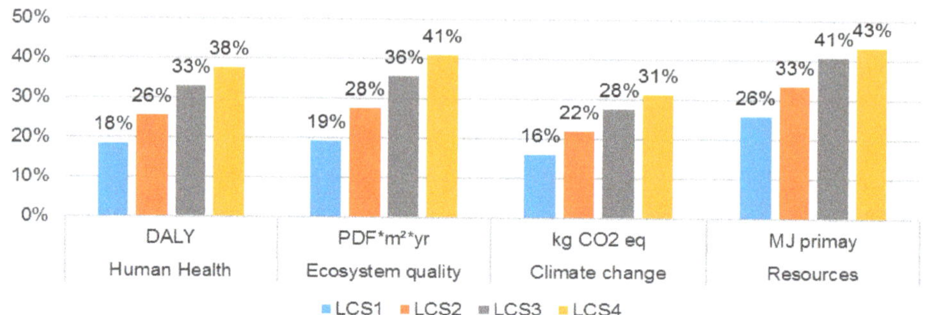

Figure 4. Relative damage category indicator compared to the baseline scenario.

The third step, normalization, was also optional according to the ISO standards. Nonetheless, it can provide valuable insight about the extent of an impact category result with a specific reference. To normalize the damage factors, each impact per unit of emission was divided by the total impact of all substances of the specific category for which characterization factors exist, per person per year (for Europe) [81]. These numbers are ratios and normalization, and therefore, resolve the mismatch of units (see Table 8).

Table 8. Total normalization results per scenario.

Damage Category	LCS0	LCS1	LCS2	LCS3	LCS4
Human Health	61.1	50	45.6	41.1	38.3
Ecosystem quality	13.7	11.1	9.95	8.82	8.1
Climate change	46.2	38.8	36.2	33.5	31.9
Resources	163	121	109	96.7	92.8

The normalization results provided a similar output to the damage assessment. Figure 5 shows the relative decrease in the normalization factor for each scenario, compared to the baseline scenario. The normalization results show great resemblance with the damage assessment outcome; the indicators of each category decrease as the RAP rate increases. For LCS4, the biggest difference was again observed in the resources category (43%) and the smallest in the climate change category (31%).

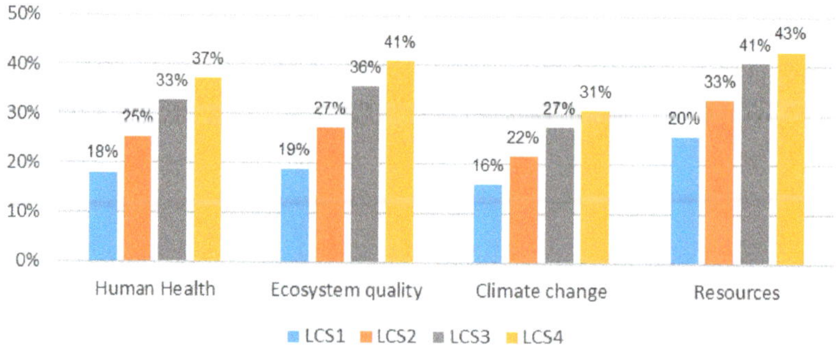

Figure 5. Normalization factor differences between LCS0 and LCS1, LCS2, LCS3 and LCS4.

A linear correlation between the RAP incorporation rate and the environmental impact can be observed. This seemed to indicate that, theoretically, when constructing or maintaining flexible pavement courses, the solutions with high RAP recycling rates should be taken into consideration.

The results from these three steps can be explained by the following aspects: (i) replacing virgin material for RAP resulted in an obvious reduction in virgin aggregate and binder production; (ii) the contribution of RAP production process were insignificant when compared with the aggregate and binder production process; (iii) multi-recycling, or in this case study, the reuse of RAP during the 4th maintenance resulted in a second reduction in virgin materials, in comparison with the baseline scenario; (iv) due to the reduction in virgin material, there was also a reduction in virgin material transport and, hence, in the overall transport impact; (v) the SMA mixtures considered, in the baseline scenario, had a higher mixing temperature than the corresponding RAP mixtures in the alternative scenarios, thus leading to a reduction in mixture production emissions and energy.

5.3. Assessment of RAP's Impacts on Mixture Production

This section only evaluates the impacts of the phases directly related with the mixture production, since RAP recycling has only direct impacts in terms of raw materials needs, RAP treatments, and additional transport distance. The following topics are intended to show the impacts of the use of this material on the overall process associated with the production of mixtures.

The LCA results from the five scenarios considered in this paper were divided by the following life cycle phases: virgin material extraction (consisting of virgin aggregate production and virgin binder production), mixture production, RAP production and transport. This makes it possible to assess the contribution of each phase to the overall environmental impact of the life cycle of each scenario; the characterization results for each category being divided by each life cycle phase. These contribution results for LCS0, LCS1, LCS2, LCS3 and LCS4 are shown in Figures 6–10, respectively.

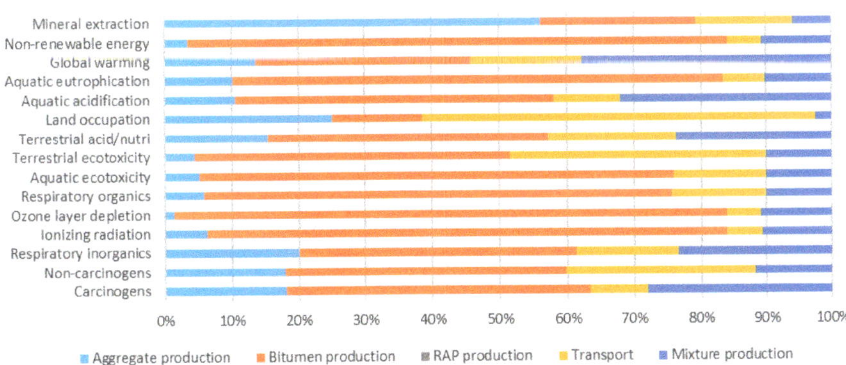

Figure 6. Life cycle phase contribution LCS0.

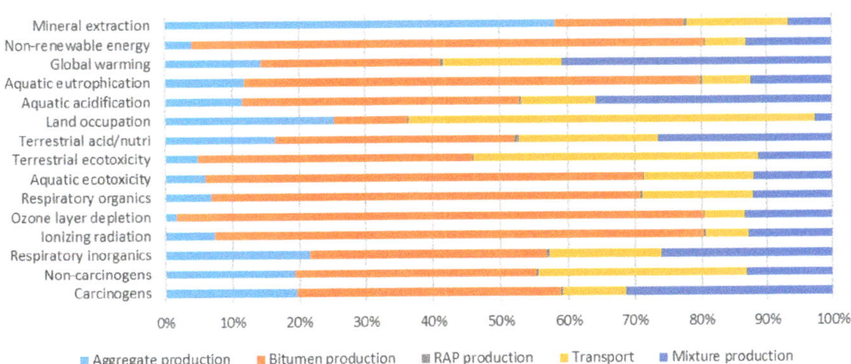

Figure 7. Life cycle phase contribution LCS1.

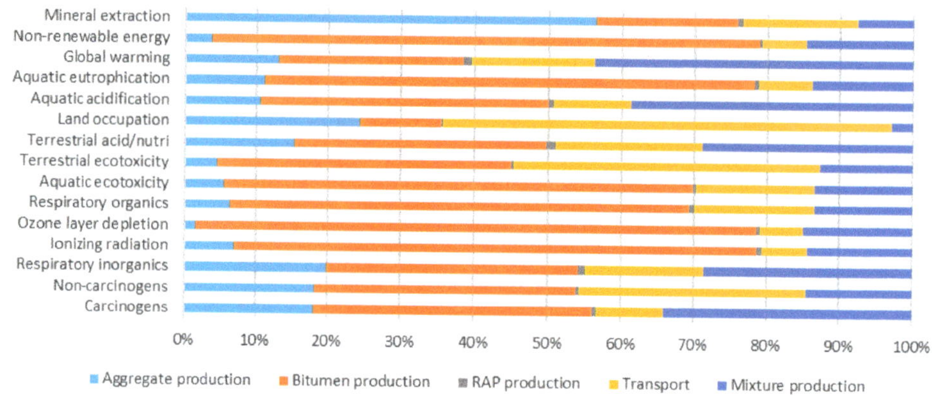

Figure 8. Life cycle phase contribution LCS2.

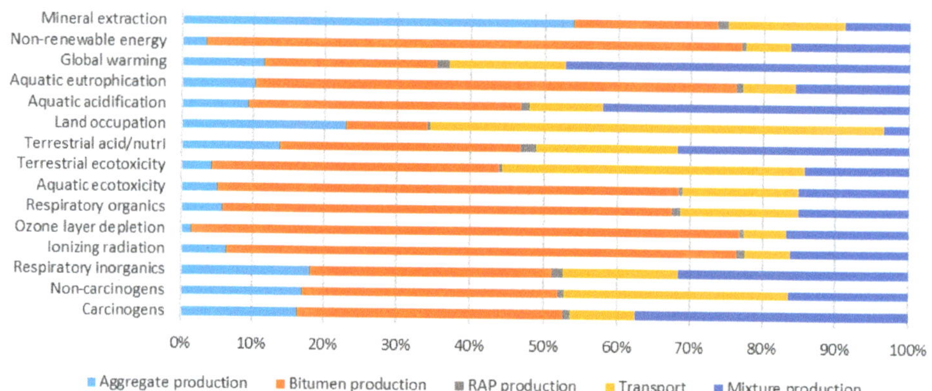

Figure 9. Life cycle phase contribution LCS3.

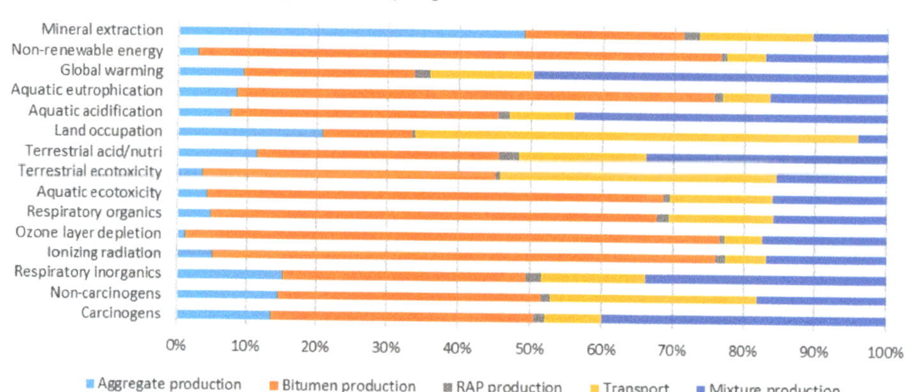

Figure 10. Life cycle phase contribution LCS4.

From these figures, it can be concluded that the production of virgin bitumen accounts for the biggest environmental impact. It has the biggest contribution to 12, 11, 11, 9 and 10 impact categories for LCS0, LCS1, LCS2, LCS3 and LCS4, respectively. This result was in line with the results obtained in other studies [11,78,82–84]. Transport and mixture production had the biggest contribution to only one, two or three categories, depending on the scenario. The production of virgin aggregate accounted for the biggest contribution to only one impact category, mineral extraction, but across all scenarios. The environmental impact of RAP production, even in LCS4, never contributed the most to any impact category.

For each rehabilitation scenario, the relative influence of the different life cycle phases was assessed in terms of the different damage categories presented in LCA global results. This provides a more comprehensible and intuitive approach to the relative contribution of each life cycle phase to the overall environmental impact; the mentioned results for each impact category are again combined into damage factors. The results obtained for each damage category are presented in Figures 11–14.

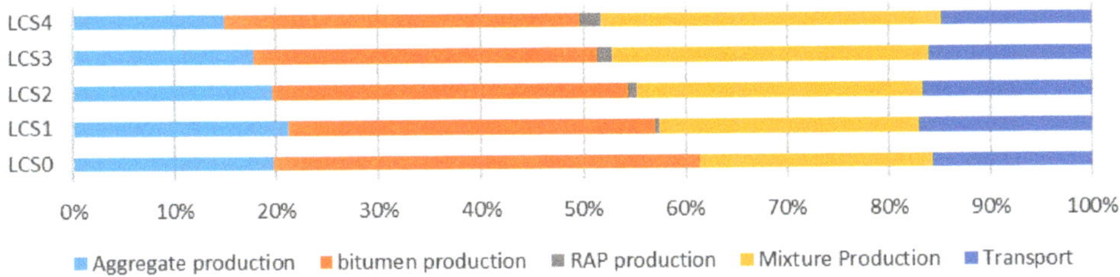

Figure 11. Relative life cycle phase contribution to the human health damage category.

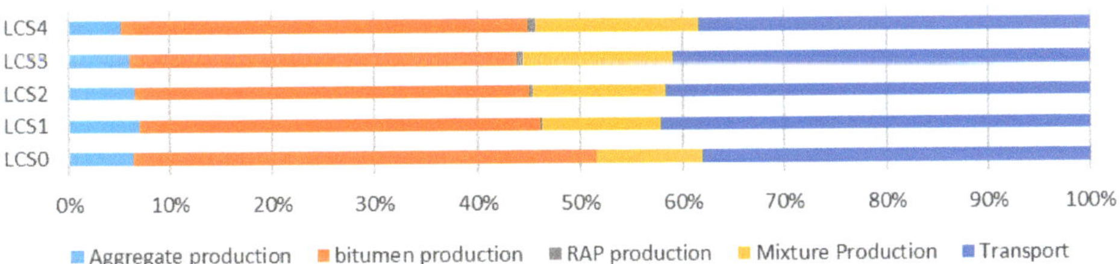

Figure 12. Relative life cycle phase contribution to the ecosystem quality damage category.

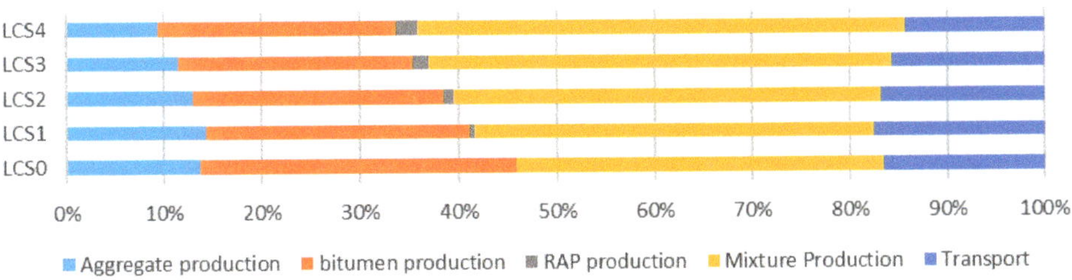

Figure 13. Relative life cycle phase contribution to the climate change damage category.

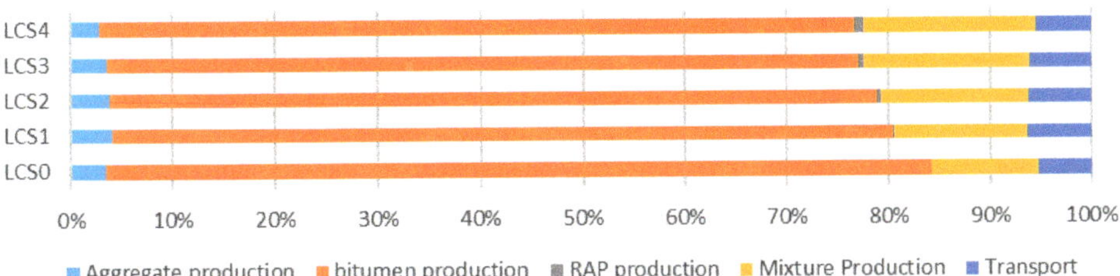

Figure 14. Relative life cycle phase contribution to the resource damage category.

Similar conclusions can be drawn from these figures in comparison with the ones showing the contribution to all impact categories. The contribution of RAP production is insignificant for all damage categories, across all scenarios. Even the relative contribution of 2% in LCS4 (100% of RAP incorporation) to human health, the largest impact observed, is still very small. On average, the production of bitumen delivers the greatest contribution across all damage categories except for climate change, in which the mixture production has the biggest impact. For the resource damage category, in particular, the contribution is almost 80% in the LCS0.

6. Conclusions

The main purpose of this paper was to evaluate the impact of high RAP incorporation and multi-recycling of bituminous mixtures in flexible pavement management. Such an impact was assessed from an environmental perspective by performing an LCA. The functional unit of this case study consisted of a flexible pavement of a road section with 1-km length and 7 m width, throughout a 69-year performance analysis period. The case study compared five maintenance and rehabilitation scenarios, referred to as 'LCSx', for this section, each with a different RAP incorporation rate in the bituminous mixtures: 0%, 25%, 50%, 75% and 100%, respectively.

The case study LCA findings were the subject of the LCIA and included characterization, damage assessment, normalization and single score results. The characterization step demonstrated the environmental impact of the five LCS life cycles across 15 impact categories. Throughout all categories, without exception, a general decrease in impacts can be observed in proportion to the RAP rate used.

The most important conclusions that can be drawn from this study are the following:

- The incorporation of RAP in bituminous mixtures, and especially considering its multi-recycling, had lesser environmental impacts than the use of solutions with only virgin materials.
- The most significant impact factors considered in this case study were human health, natural resources and climate change or global warming. The LCA case study demonstrated that across all factors, RAP incorporation is beneficial and significantly reduces the impact.
- Moreover, a linear correlation was established between the RAP incorporation rate and the environmental impact. This suggested that, theoretically, when constructing or maintaining flexible pavement courses, a 100% RAP incorporation (or the complete recycling and multi-recycling of materials) should be pursued.

Although these results may seem promising, it should be considered that this case study represented certain practices, and hence it should be adapted for other construction, maintenance and rehabilitation scenarios, and for different geographic regions or types of pavements.

It should be pointed out that the data for this study was obtained from multiple active companies and experts to provide the most possible accurate and comprehensive input for the case study.

In conclusion, the case study of a road section rehabilitation has demonstrated a clear decrease in the environmental impact when RAP was reused, either once or multiple times, during M&R phases. The multi-recycling of flexible pavement should, therefore, be favored over using virgin materials in new bituminous mixtures. Besides the study on the environmental benefits of RAP multi-recycling, significant worldwide research has been carried out to address the mechanical and durability advantages of this technology with promising results.

Bearing in mind that decision-making in pavement management is a fairly difficult approach, the LCA approach should be integrated toward a multi-purpose optimization framework and should include factors such as: social dimension, structural objectives, constraints and life cycle cost analyses. In this way, it should be performed to calculate the associated costs. Furthermore, it should be noted that the choice for the software tools to perform an LCA, as well as for the database and the impact assessment method has a distinctive influence on results. Therefore, further work should be done with a view to cross-examine the same functional unit across different tools.

Author Contributions: Conceptualization, V.A.; methodology, V.A.; software, D.V.; validation, V.A.; formal analysis, D.V. and V.A.; investigation, D.V.; resources, J.N., A.C.F. and V.A.; writing—original draft preparation, D.V.; writing—review and editing, V.A., J.N. and A.C.F.; visualization, D.V.; supervision, J.N. and A.C.F. All authors have read and agreed to the published version of the manuscript.

Funding: This research was funded by Portuguese Foundation for Science and Technology, IP, grant number SFRH/BD/114715/2016.

Acknowledgments: The authors are grateful to the Portuguese Foundation for Science and Technology, IP, for the financial support provided through grant SFRH/BD/114715/2016 financed by the Portuguese Government budget. The authors also want to express their gratitude to Nele De Belie from the University of Ghent for providing the license for software SimaPro. The authors are grateful to the companies that provided the data.

Conflicts of Interest: The authors declare no conflict of interest.

References

1. NAPA; EAPA. *The Asphalt Paving Industry—A Global Perspective*, 2nd ed.; NAPA: Brussels, Belgium; EAPA: Lanham, MD, USA, 2011.

2. CEN. *EN 13108-8: Bituminous Mixtures—Material Specifications—Part 8: Reclaimed Asphalt*; European Committee for Standardization: Brussels, Belgium, 2005.

3. European Commission Sustainability and Circular Economy—Circular Economy. 2018. Available online: https://ec.europa.eu/growth/industry/sustainability/circular-economy_en (accessed on 17 June 2018).

4. COM 21 A Resource-Efficient Europe—Flagship Initiative under the Europe 2020 Strategy. *Communication from the Commision to the European Parliament, the Council, the European Economic and Social Committee and the Committee of the Regions.* 2011. Available online: https://eur-lex.europa.eu/legal-content/en/TXT/?uri=CELEX%3A52011DC0021 (accessed on 5 April 2020).

5. FIEC; EBC; EFBWW; UIPI; Eurima; BUILD EUROPE; RICS; CEMBUREAU; GCP EUROPE, EDA; et al. CONSTRUCTION 2050—Building Tomorrow's Europe Today. 2019. Available online: http://www.fiec.eu/en/news/news-2019/construction-2050-building-tomorrows-europe-today.aspx (accessed on 15 March 2020).

6. Santos, J.; Cerezo, V.; Flintsch, G.; Ferreira, A. Environmental and economic assessment of pavement construction and management practices for enhancing pavement sustainability. *Resour. Conserv. Recycl.* **2017**, *116*, 15–31. [CrossRef]

7. EAPA. *Asphalt the 100% Recyclable Construction Product EAPA Position Paper*; EAPA: Brussels, Belgium, 2014.

8. EAPA; EUPAVE; FEHRL. Road Pavement Industries Highlight Huge CO2 Savings Offered by Maintaining and Upgrading Roads. 2016. Available online: https://eapa.org/eapa-eupave-fehrl-position-paper/ (accessed on 15 March 2020).

9. Said, F.M.; Bolong, N.; Gungat, L. Life Cycle Assessment of Asphalt Pavement Construction and Maintenance—A Review. In Proceedings of the 10th Annual Seminar on Science and Technology, Pahang, Malaysia, 1–2 December 2012.

10. Anthonissen, J.; Van den bergh, W.; Braet, J. Review and environmental impact assessment of green technologies for base courses in bituminous pavements. *Environ. Impact Assess. Rev.* **2016**, *60*, 139–147. [CrossRef]

11. Farina, A.; Zanetti, M.C.; Santagata, E.; Blengini, G.A. Life cycle assessment applied to bituminous mixtures containing recycled materials: Crumb rubber and reclaimed asphalt pavement. *Resour. Conserv. Recycl.* **2017**, *117*, 204–212. [CrossRef]

12. Butt, A.A.; Toller, S.; Birgisson, B. Life cycle assessment for the green procurement of roads: A way forward. *J. Clean. Prod.* **2015**, *90*, 163–170. [CrossRef]

13. Keijzer, E.E.; Leegwater, G.A.; de Vos-Effting, S.E.; de Wit, M.S. Carbon footprint comparison of innovative techniques in the construction and maintenance of road infrastructure in The Netherlands. *Environ. Sci. Policy* **2015**, *54*, 218–225. [CrossRef]

14. Hunt, R.G.; Franklin, W.E. LCA History L C A—How it Came About—Personal Reflections on the Origin and the Development of LCA in the USA. *Int. J. Life Cycle Assess.* **1996**, *1*, 4–7. [CrossRef]

15. Butt, A.A.; Mirzadeh, I.; Toller, S.; Birgisson, B. Life cycle assessment framework for asphalt pavements: Methods to calculate and allocate energy of binder and additives. *Int. J. Pavement Eng.* **2014**, *15*, 290–302. [CrossRef]

16. Jiang, R.; Wu, P. Estimation of environmental impacts of roads through life cycle assessment: A critical review and future directions. *Transp. Res. Part D Transp. Environ.* **2019**, *77*, 148–163. [CrossRef]

17. CEN. *Environmental Management—Life Cycle Assessment—Principles and Framework*; ISO EN 14040; European Committee for Standardization: Brussels, Belgium, 2006.

18. Steen, B.; Carlson, R.; Lyrstedt, F.; Skantze, G. *Sustainability Management of Businesses through Eco-Efficiency—An EXAMPLE*; CPM Report No. 2009:3; Chalmers University of Technology: Göteborg, Sweden, 2009.

19. Wolf, M.-A.; Pant, R.; Chomkhamsri, K.; Sala, S.; Pennington, D. *The International Reference Life Cycle Data System (ILCD) Handbook: Towards More Sustainable Production and Consumption for a Resource-Efficient Europe*; European Commision: Brussels, Belgium, 2012; ISBN 9789279216404.

20. Emami, N.; Heinonen, J.; Marteinsson, B.; Säynäjoki, A.; Junnonen, J.M.; Laine, J.; Junnila, S. A life cycle assessment of two residential buildings using two different LCA database-software combinations: Recognizing uniformities and inconsistencies. *Buildings* **2019**, *9*, 10020. [CrossRef]

21. West, R.; Rodezno, C.; Julian, G.; Prowell, B.; Frank, B.; Kriech, T. *NCHRP Report 779. Field Performance of Warm Mix Technologies*; Transportation Research Board: Washington, DC, USA, 2014. [CrossRef]

22. Pennington, D.W.; Margni, M.; Ammann, C.; Jolliet, O. Multimedia fate and human intake modeling: Spatial versus nonspatial insights for chemical emissions in Western Europe. *Environ. Sci. Technol.* **2005**, *39*, 1119–1128. [CrossRef]

23. IPCC. 2001: Climate Change 2001: The Scientific Basis. In *Contribution of Working Group I to the Third Assessment Report of The Intergovernmental Panel on Climate Change*; Houghton, J.T., Ding, Y., Griggs, D.J., Noguer, M., van der Liden, P.J., Dai, X., Maskell, K., Johnson, C.A., Eds.; Cambridge University Press: Cambrige, UK; New York, NY, USA, 2001; p. 881.

24. Araújo, J.P.C.; Oliveira, J.R.M.; Silva, H.M.R.D. The importance of the use phase on the LCA of environmentally friendly solutions for asphalt road pavements. *Transp. Res. Part D Transp. Environ.* **2014**, *32*, 97–110. [CrossRef]

25. CEN. *Tests for General Properties of Aggregates—Part 1: Methods for Sampling*; NP EN 932-1:20022002; British Standard Institution: London, UK, 1997.

26. Cameron, D.A.; Azam, A.H.; Rahman, M.M. Recycled clay masonry and recycled concrete aggregate blends in pavement. In Proceedings of the GeoCongress 2012: State of the Art and Practice in Geotechnical Engineering, Oakland, CA, USA, 25–29 March 2012; pp. 1532–1541. [CrossRef]

27. Rahman, M.M.; Beecham, S.; Iqbal, A.; Karim, M.R.; Rabbi, A.T.Z. Sustainability assessment of using recycled aggregates in concrete block pavements. *Sustainability* **2020**, *12*, 4314. [CrossRef]

28. Europen Commission. 614—Communication from the Commission to the European Parliament, the Council, The European Economic and Social Committee and the Committee of the Regions. In *Closing the Loop—An EU Action Plan for the Circular Economy*; European Commission: Brussels, Belgium, 2015.

29. Antunes, M.D.L. Circular economy: Construction with added value (in Portuguese). In Proceedings of the Seminar: CDW a Valuable Resource (in Portuguese), Lisbon, Portugal, 19 September 2016.

30. Macarthur, F.E. Circular Economy. 2011. Available online: https://www.ellenmacarthurfoundation.org/circular-economy/infographic (accessed on 22 October 2018).

31. Zaumanis, M.; Mallick, R.B.; Frank, R. Use of Rejuvenators for Production of Sustainable High Content Rap Hot Mix Asphalt. In Proceedings of the XXVIII International Baltic Road Conference, Vilnius, Lithuania, 26–28 August 2013; pp. 1–10.

32. Zaumanis, M.; Mallick, R.B.; Frank, R. 100% recycled hot mix asphalt: A review and analysis. *Resour. Conserv. Recycl.* **2014**, *92*, 230–245. [CrossRef]

33. Silva, H.M.R.D.; Oliveira, J.R.M.; Jesus, C.M.G. Are totally recycled hot mix asphalts a sustainable alternative for road paving? *Resour. Conserv. Recycl.* **2012**, *60*, 38–48. [CrossRef]

34. Martinho, F.C.G.; Picado-santos, L.G.; Capitão, S.D. Mechanical properties of warm-mix asphalt concrete containing different additives and recycled asphalt as constituents applied in real production conditions. *Constr. Build. Mater.* **2017**, *131*, 78–89. [CrossRef]

35. Porot, L.; Di Nolfo, M.; Polastro, E.; Tulcinsky, S. Life cycle evaluation for reusing Reclaimed Asphalt with a bio-rejuvenating agent. In Proceedings of the 6th Eurasphalt & Eurobitume Congress, Prague, Czech Republic, 1–3 June 2016; pp. 1–8. [CrossRef]

36. McDaniel, R.; Michael Anderson, R. *Recommended Use of Reclaimed Asphalt Pavement in the Superpave Mix Design Method: Technician's Manual Transportation*; NCHRP REPORT 452; National Academy Press: Washington, DC, USA, 2001.

37. Antunes, V.; Freire, A.C.; Neves, J. A review on the effect of RAP recycling on bituminous mixtures properties and the viability of multi-recycling. *Constr. Build. Mater.* **2019**, *211*, 453–469. [CrossRef]
38. Al-Qadi, I.L.; Aurangzeb, Q.; Carpenter, S.H.; Pine, W.J.; Trepanier, J. *Impact of High RAP Content on Structural and Performance Properties of Asphalt Mixtures*; FHWA-ICT-12-002; Illinois Center for Transportation: Rantoul, IL, USA, 2012.
39. Batouli, M.; Bienvenu, M.; Mostafavi, A. Putting sustainability theory into roadway design practice: Implementation of LCA and LCCA analysis for pavement type selection in real world decision making. *Transp. Res. Part D Transp. Environ.* **2017**, *52*, 289–302. [CrossRef]
40. Zhang, Y.; Goulias, D.; Aydilek, A. Sustainability evaluation of pavements using recycled materials. In *Bearing Capacity of Roads, Railways and Airfields, Proceedings of the 10th International Conference on the Bearing Capacity of Roads, Railways and Airfields—BCRRA 2017, Athens, Greece, 28–30 June 2017*; Loizos, A., Al-Qadi, I., Scarpas, T., Eds.; CRC Press—Taylor & Francis Group: London, UK, 2017; pp. 1283–1291. [CrossRef]
41. Zaumanis, M.; Mallick, R.B.; Frank, R. 100% Hot Mix Asphalt Recycling: Challenges and Benefits. *Transp. Res. Procedia* **2016**, *14*, 3493–3502. [CrossRef]
42. West, R.; Willis, J.R.; Marasteanu, M. *Improved Mix Design, Evaluation, and Materials Management Practices for Hot Mix Asphalt with High Reclaimed Asphalt Pavement Content*; NCHRP Report 752; National Academy Press: Washington, DC, USA, 2013.
43. Elkashef, M.; Williams, R.C. Improving fatigue and low temperature performance of 100% RAP mixtures using a soybean-derived rejuvenator. *Constr. Build. Mater.* **2017**, *151*, 345–352. [CrossRef]
44. Huang, B.; Zhang, Z.; Kingery, W.; Zuo, G. Fatigue crack characteristics of HMA mixtures containing RAP. In Proceedings of the Fifth International RILEM Conference on Reflective Cracking in Pavements, Limoges, France, 5–8 May 2004; pp. 631–638.
45. Nguyen, V.H. Effects of Laboratory Mixing Methods and RAP Materials on Performance of Hot Recycled Asphalt Mixtures. Ph.D. Thesis, University of Nottingham, Nottingham, UK, 2009.
46. Mogawer, W.S.; Booshehrian, A.; Vahidi, S.; Austerman, A.J. Evaluating the effect of rejuvenators on the degree of blending and performance of high RAP, RAS, and RAP/RAS mixtures. *Road Mater. Pavement Des.* **2013**, *14*, 193–213. [CrossRef]
47. Baptista, A.M.C. Hot Recycled Bituminous Mixtures in Plant—A Contribution to Its Study and application. Ph.D. Thesis, University of Coimbra, Coimbra, Portugal, 2006. (In Portuguese).
48. Al-Qadi, I.L.; Carpenter, S.H.; Roberts, G.; Ozer, H.; Aurangzeb, Q.; Elseifi, M.; Trepanier, J. *Determination of Usable Residual Asphalt Binder in RAP*; Research Report ICT-09-031; Illinois Center for Transportation: Rantoul, IL, USA, 2009.
49. Henely, R. Evaluation of Recycled Asphalt Concrete Pavements—Project SN-1179(6). Iowa. 1980. Available online: https://trid.trb.org/view/1124675 (accessed on 13 August 2020).
50. Hellriegel, E.J. *FHWA/NJ-81/002—Bituminous Concrete Pavement Recycling*; Federal Highway Administration: Washington, DC, USA, 1980.
51. Santos, L.G.d.P.; Baptista, A.M.d.C.; Capitão, S.D. Assessment of the Use of Hot-Mix Recycled Asphalt Concrete in Plant. *J. Transp. Eng. ASCE* **2010**, *136*, 1159–1164. [CrossRef]
52. Baptista, A.M.; Picado-Santos, L.G.; Capitão, S.D. Design of hot-mix recycled asphalt concrete produced in plant without preheating the reclaimed material. *Int. J. Pavement Eng.* **2013**, *14*, 95–102. [CrossRef]
53. Bloomquist, D.; Diamond, G.; Oden, M.; Ruth, B.; Tia, M. *Engineering and Environmental Aspects of Recycled Materials for Highway Construction*; FHWA-RD-93-088; Western Research Inst.: Laramie, WY, USA, 1993.
54. Little, D.N.; Epps, J.A. Evaluation of Certain Structural Characteristics of Recycled Pavement Materials. In Proceedings of the Association of Asphalt Paving Technologists Proceedings, Louisville, KY, USA, 18–20 February 1980.
55. Sullivan, J. *FHWA-SA-95-060. Pavement Recycling Executive Summary and Report*; Federal Highway Administration: Washington, DC, USA, 1996.
56. Zaumanis, M.; Mallick, R.B.; Frank, R. Evaluation of Rejuvenator's Effectiveness with Conventional Mix Testing for 100% RAP Mixtures. In Proceedings of the Transportation Research Board 92nd Annual Meeting, Washington, DC, USA, 13–17 January 2013; pp. 17–25.
57. Pereira, P.A.A.; Oliveira, J.R.M.; Picado-Santos, L.G. Mechanical characterisation of hot mix recycled materials. *Int. J. Pavement Eng.* **2004**, *5*, 211–220. [CrossRef]

58. Valdés, G.; Pérez-Jiménez, F.; Miró, R.; Martínez, A.; Botella, R. Experimental study of recycled asphalt mixtures with high percentages of reclaimed asphalt pavement (RAP). *Constr. Build. Mater.* **2011**, *25*, 1289–1297. [CrossRef]

59. Shu, X.; Huang, B.; Vukosavljevic, D. Laboratory evaluation of fatigue characteristics of recycled asphalt mixture. *Constr. Build. Mater.* **2008**, *22*, 1323–1330. [CrossRef]

60. Daniel, J.S.; Lachance, A. Mechanistic and volumetric properties of asphalt mixtures with recycled asphalt pavement. *Bitum. Paving Mix. 2005* **2005**, 28–36. [CrossRef]

61. Botella, R.; Pérez-Jiménez, F.; Miró, R.; Guisado-Mateo, F.; Ramírez Rodríguez, A. Characterization of Half-Warm–Mix Asphalt with High Rates of Reclaimed Asphalt Pavement. *Transp. Res. Rec. J. Transp. Res. Board* **2016**, *2575*, 168–174. [CrossRef]

62. Farooq, M.A.; Mir, M.S.; Sharma, A. Laboratory study on use of RAP in WMA pavements using rejuvenator. *Constr. Build. Mater.* **2018**, *168*, 61–72. [CrossRef]

63. Wang, Y. The effects of using reclaimed asphalt pavements (RAP) on the long-term performance of asphalt concrete overlays. *Constr. Build. Mater.* **2016**, *120*, 335–348. [CrossRef]

64. Stroup-Gardiner, M. *Use of Reclaimed Asphalt Pavement and Recycled Asphalt Shingles in Asphalt Mixtures*; Transportation Research Board: Washington, DC, USA, 2016.

65. Gong, H.; Huang, B.; Shu, X. Field performance evaluation of asphalt mixtures containing high percentage of RAP using LTPP data. *Constr. Build. Mater.* **2018**, *176*, 118–128. [CrossRef]

66. Ecoinvent System Models in Ecoinvent 3. 2019. Available online: https://www.ecoinvent.org/database/system-models-in-ecoinvent-3/system-models-in-ecoinvent-3.html (accessed on 6 September 2019).

67. Santos, J.; Bressi, S.; Cerezo, V.; Lo Presti, D.; Dauvergne, M. Life cycle assessment of low temperature asphalt mixtures for road pavement surfaces: A comparative analysis. *Resour. Conserv. Recycl.* **2018**, *138*, 283–297. [CrossRef]

68. Rodriguez-Alloza, A.M.; Malik, A.; Lenzen, M.; Gallego, J. Hybrid input-output life cycle assessment of warm mix asphalt mixtures. *J. Clean. Prod.* **2015**, *90*, 171–182. [CrossRef]

69. Bressi, S.; D'Angelo, G.; Santos, J.; Giunta, M. Environmental performance analysis of bitumen stabilized ballast for railway track-bed using life-cycle assessment. *Constr. Build. Mater.* **2018**, *188*, 1050–1064. [CrossRef]

70. Walls-III, J.; Smith, M.R. *Life-Cycle Cost Analysis in Pavement Design—Interim Technical Bulletin*; FHWA-SA-98-079; Federal Highway Administration: Washington, DC, USA, 1998.

71. Babashamsi, P.; Md Yusoff, N.I.; Ceylan, H.; Md Nor, N.G.; Salarzadeh Jenatabadi, H. Evaluation of pavement life cycle cost analysis: Review and analysis. *Int. J. Pavement Res. Technol.* **2016**, *9*, 241–254. [CrossRef]

72. Thiessen, P.; Collins, J.; Buckland, T.; Abbell, R. Valuing the wider benefits of road maintenance funding. *Transp. Res. Procedia* **2017**, *26*, 156–165. [CrossRef]

73. Dinis-Almeida, M.; Castro-Gomes, J.; Antunes, M.L. Mechanical Performance and Economic Evaluation of Warm Mix Recycling Asphalt. *Procedia Soc. Behav. Sci.* **2012**, *53*, 286–296. [CrossRef]

74. JRC-IES. *International Reference Life Cycle Data System (ILCD) Handbook—General Guide for Life Cycle Assessment—Detailed Guidance*; European Commission—Joint Research Centre—Institute for Environment and Sustainabilitty: Luxemborg, 2010.

75. Pré Consultants. SimaPro Database Manual Methods Library. 2019. Available online: http://creativecommons.org/licenses/by-nc-sa/3.0/nl/deed.en_US (accessed on 15 February 2019).

76. Pré Consultants. SimaPro 7—Database Manual—EU & DK Input Output Database. 2010. Available online: https://www.pre-sustainability.com/legacy/download/manuals/DatabaseManualEU-DKIODatabase.pdf (accessed on 15 February 2019).

77. CEN. *Bituminous Mixtures. Test Methods for Hot Mix Asphalt—Part 35: Laboratory Mixing*; EN 12697-35; European Committee for Standardization: Brussels, Belgium, 2004.

78. Santero, N.J.; Masanet, E.; Horvath, A. Life-cycle assessment of pavements. Part I: Critical review. *Resour. Conserv. Recycl.* **2011**, *55*, 801–809. [CrossRef]

79. GocEkoop, M.; Spriemsma, R. *The Eco-Indicator 99 A Damage Oriented Method for Life Cycle Impact Assessment*, 3rd ed.; PRé Consultants B.V.: Amersfoort, The Netherlands, 2001.

80. Guinée, J. *Handbook on Life Cycle Assessment: Operational Guide to the ISO Standards (Eco-Efficiency in Industry and Science)*; Kluwer Academic Publishers: Dordrecht, The Netherlands, 2002; ISBN 978-1402005572.

81. Humbert, S.; De Schryver, A.; Margni, M.; Jolliet, O. IMPACT 2002+: User Guide. 2012. Available online: http://onlinelibrary.wiley.com/o/cochrane/clcentral/articles/506/CN-00756506/frame.html (accessed on 20 February 2019).
82. Bressi, S.; Santos, J.; Giunta, M.; Pistonesi, L.; Lo Presti, D. A comparative life-cycle assessment of asphalt mixtures for railway sub-ballast containing alternative materials. *Resour. Conserv. Recycl.* **2018**, *137*, 76–88. [CrossRef]
83. Santos, J.; Ferreira, A.; Flintsch, G. A life cycle assessment model for pavement management: Methodology and computational framework. *Int. J. Pavement Eng.* **2015**, *16*, 268–286. [CrossRef]
84. Santos, J. A Comprehensive Life Cycle Approach for Managing Pavement Systems. Ph.D. Thesis, University of Coimbra, Coimbra, Portugal, 2015.

MDPI
St. Alban-Anlage 66
4052 Basel
Switzerland
Tel. +41 61 683 77 34
Fax +41 61 302 89 18
www.mdpi.com

Recycling Editorial Office
E-mail: recycling@mdpi.com
www.mdpi.com/journal/recycling